OWNERSHIP STABLING & FEEDING

HORSEKEEPING

OWNERSHIP STABLING & FEEDING

Ray Saunders

Frederick Muller Limited
London

General Editor: Candida Hunt

In preparation in the same illustrated series
Horsekeeping–Management and Ailments
Horsekeeping–Handling and Training
Horsekeeping–Small Scale Breeding

First Published in Great Britain in 1982
by Frederick Muller Limited,
Dataday House, London SW19 7JU

Produced for the publisher
by Midas Books
12 Dene Way, Speldhurst,
Tunbridge Wells, Kent TN3 0NX

Copyright © Ray Saunders 1982

All rights reserved. No part of this publication may be reproduced, stored in a retrieval system, or transmitted, in any form or by any means, electronic, mechanical, photocopying, recording or otherwise, without the prior consent of Midas Books.

ISBN 0 584 95009 8 Hardback

ISBN 0 584 95010 1 Softcover

Printed by Nene Litho.
Bound by Woolnough Bookbinding,
Wellingborough, Northants.

Contents

		page
1	Owning your first horse or pony	1
2	Stabling facilities	19
3	Managing your stable	35
4	Keeping a horse or pony at grass	45
5	Feeding requirements	59
6	Tack and other equipment	69
7	Stable routine and discipline	83
8	Care of the feet	93
	Index	105
	Colour plate section	between pages 58-59

Acknowledgements

My grateful thanks to Jack Neal for working with me to produce all the photographs for this book. His patience and expertise in filming, and his technique in developing and printing both colour and black and white, have ensured that each photograph shows exactly what is required. The cameras used were an Olympus OM2 with 35 mm wide-angle, 50 mm standard and 210 mm zoom lenses, plus macro, together with a Mamia C330 reflex (2¼ square). Films used were Ilford Pan F and Ektachrome 64.

My thanks also to my editor, Candida Hunt, for making many useful suggestions that enabled the text to be made more informative as well as concise.

1 Owning your first horse or pony

There can be few thrills greater than that of owning your very own horse or pony. For a few specially talented youngsters bitten by the horse bug it may be the first step to a brilliant international career; at the other end of the scale a lifetime's ambition may be realised by owning one's own quiet hack to enjoy in retirement. Whether you are a child or an adult, and no matter how many horses you eventually own, nothing quite compares with the thrill of getting your first one.

Horses and ponies are without doubt the most stupid, frustrating, disappointing and time-consuming animals; they are also the most enjoyable and rewarding. This is especially true if the first one proves to be a 'good un', which makes the choice of your first horse particularly important.

It may be that you are interested in owning a horse that you can use mainly for hacking, with the additional pleasure of some dressage training at your local riding school and perhaps the occasional excitement of some novice eventing. For those lucky enough to live in parts of Britain where hills and open moorlands still exist, and for those in countries such as the United States and Australia where there are large areas of plains and prairies, a mount that is tough enough to cope with the local country and conditions will be required. This sounds obvious enough, but it is surprising how easy it is to be misled or talked into acquiring the wrong sort of animal. It is also important to be realistic about your own riding standard, as many a good horse has been ruined through no other fault than the rider's lack of ability.

The purchase of a horse or pony is not something to be rushed into, and some old but sound advice on paying as much attention to the person selling the horse as you do to the horse itself doesn't

Good examples of the types of ponies used for riding club and Pony Club activities.

come amiss at this point. You may be shown an old dropdead that has been corned up and 'doctored' so as to put on a once-in-a-lifetime performance for your benefit, or alternatively a tearaway who has been kept short of water and subjected to other tricks of the trade to make him well-behaved during an inspection, but if bought would result in your being considerably overmounted.

After deciding what you will want to do with your horse, the next step is to look around at what is available and the prices being asked. Obviously you will want the best animal you can get and will be looking for perfection. Remember, though, that the perfect horse does not exist – at least I have never seen one – and some allowance will have to be made for minor imperfections or you will never buy anything. The more nearly perfect the horse is the more it will cost, so you will have to decide how much the animal is worth to you for the purpose that you wish to use it, and whether its faults are of minor or major significance for that purpose.

In Britain there are fortunately many breeds and types to choose from, particularly pony breeds suitable for children of all ages and sizes. Many other countries also have excellent native breeds, many of which have been improved by the addition of Thoroughbred, Arab or Spanish blood during their evolution. With larger horses, readers of this book will be best advised to refrain from considering anything with too much Thoroughbred blood. The Thoroughbred is by nature a highly-strung animal, having been bred for many years to produce an abundance of explosive energy. The more 'blood' an animal has in its veins, the less likely it is to be easily controlled, and the more likely it is to be lacking in what I consider the most important characteristic in a pleasure horse, namely temperament.

Do not be persuaded to buy a racing-type animal because it looks good and is reasonably priced; beware of anything too leggy, and remember that a Thoroughbred weed, no matter how cheap, is never a bargain. Beauty is as beauty does, and a more common type possessing an amiable character will almost always serve you better and be a happier purchase.

Conformation

For the purpose of choosing your own horse you will need to know something about the subject of conformation generally. Unfortunately, this is not something that can be learned merely by reading a book on the subject, as a great deal of practical experience is necessary to recognise the various points.

Unless you are very knowledgeable, it would be wise to get the advice of a vet or of friends who already own horses, or both, to assist you in making your choice. However, I will explain some of the things that are considered good and bad in the conformation of the horse. By studying them you will be able to understand what is being talked about by the experts. Visiting shows, in order to compare the winners with the others in each class, should also help you to build up a picture of what to look for when you buy. With some practice you will also soon be able to spot the most obvious faults, and thereby save yourself the expense of getting a vet to look at an obviously unsuitable animal. Vets do not come cheap, and a horse vet will probably have to travel a considerable distance to get to you. This involves his time and travelling expenses as well as the cost of his advice. It is not the same as popping the family pet along to the vet in town.

Once you have found a likely prospect that appears to satisfy you in general appearance, is of reasonable proportions and seems good-natured, a more detailed inspection will need to follow.

The first consideration is whether the animal has sufficient 'bone' for your requirements. The term bone is given to the measurement round the foreleg, taken just below the knee. Although the quality

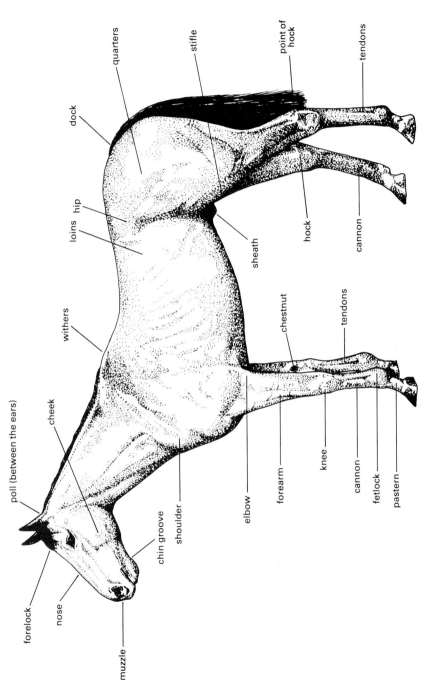

Fig. 1 The names given to the various parts of the horse, often known as the 'points'.

A strong pony of the type used for hunting.

of the bone is important (this can only be ensured by proper early feeding and the tendencies of the breed) the actual measurement is generally used to gauge robustness and weight-carrying capabilities. The 'bone' of the horse will vary according to its height and type: the taller a horse stands in hands high (a hand being 4 inches or 10 cm) the greater this measurement should be, and it should also be larger in horses and ponies of a cobby type.

Adequate bone is one important factor in an animal that one intends to use for tough and demanding riding and in conditions that are likely to play havoc with a horse's legs. Uneven stony going, especially when combined with hills and steep descents, will place a great strain on this part of the animal's anatomy. The measurement includes both the cannon bone and the tendons directly behind it. One cannot be too specific about what the actual measurement should be, as it varies with the size and type of animal. However, it is worth noting that many hunter types are proudly advertised as having 9 inches (22.8 cm) of bone and even more. This amount of bone in a hunter would indicate that it should be capable of carrying 14 st (89 kg) or more, whereas one with 8 inches (20.3 cm) below the knee would probably be judged as being up to about 12 st (76 kg). Horses lacking the required amount of bone for their height are referred to as being 'light' or 'light of bone'.

When the circumference of the cannon bone is less just below the knee than it is lower down, the term 'tied in below the knee' is used. This is considered a bad fault, as it usually indicates weak bone and

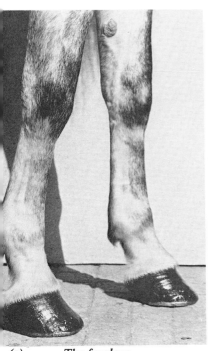

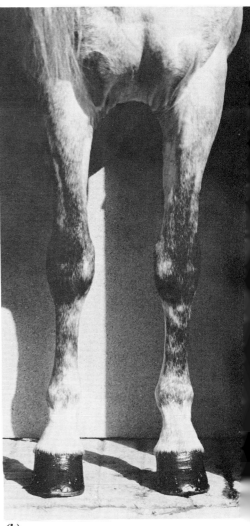

The forelegs.
(a) Lateral view showing straight clean legs with a good span of bone, short cannons and strong tendons. The angle of foot and pastern is neither too sloping nor too upright.
(b) Anterior view. The limbs are straight and parallel to each other, with a chest space neither too narrow nor too wide; the knees are broad and flat, the fetlocks and pasterns clean-looking with toes pointing forwards – not pigeon-toed or turned out.

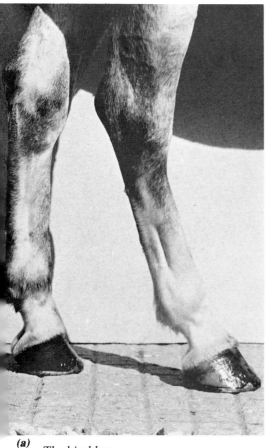
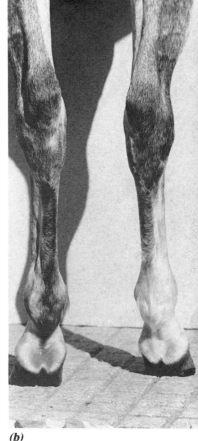

(a) The hind legs.
(a) Lateral view of normal hind legs with good, well-defined points to the hocks and showing strong cannons and tendons.
(b) Posterior view. The hocks are strong and well-matched; the tendons are clean and strong and the feet have good heels.

tendons. The cannon bone itself should be short in length and have strong-looking, clearly defined tendons. If you run a hand down them they should feel 'clean', without lumps or irregularities, and there should be no fleshiness in this region. It is also better for them to feel flat as opposed to round, the former suggesting more strength of bone.

The fetlock joint, located below the cannon bone, should also feel firm, with flat sides rather than round fleshy ones. Below this the pasterns run down into the hoof. How sloping or upright the resulting angle should be – by which I mean the amount that the hoof is placed forward of the fetlock joint – gives rise to considerable argument among horsemen. This part of the leg acts as a shock-absorber for the movement of the horse, and if the pasterns are short and upright the shock-absorber action is reduced and the legs are subjected to more jarring. Long pasterns, with excessive forward inclination, provide great springiness and result in a very smooth ride, but because of the additional strain placed on the tendons often indicate potential weakness. The two extremes are therefore best avoided and something midway between should be looked for. On most horses and ponies the hind pasterns will be slightly shorter than those in front, because of the compensating action of the hock.

Hocks probably give rise to more conflicting opinions regarding their conformation than any other part of the horse. Many consider that the perfect hock should be low-set, having its point on a horizontal level with the chestnut above the knee of the foreleg. An imaginary line is then drawn from the point of the buttock, passing down the rear point of the hock, to continue in a straight line down the rear of the cannon bone to the ground. Theoretically, this is said to represent perfect conformation. In practice many horses deviate from this with no ill effect. Sometimes the hocks stand behind this imaginary line, but it is more common for them to stand forward of it. If in addition to this the cannon bone slopes forward from the point of the hock to the ground, the horse is said to have 'sickle hocks'. The hocks then have an overbent, curved appearance on the front surface. Most authorities consider this a weakness; I do not believe this is necessarily so, provided it is not excessively pronounced.

Other divergences from the 'correct' placement are when the hocks bow outwards so that the points are carried far apart when viewed from the rear, or when they bend inwards so that the points come close together. Neither of these two conditions is to be welcomed, the former being known as 'bowed hocks' and the latter 'cow hocks'. It is true to say, however, that many horses shaped in this way are still able to carry out their duties satisfactorily. More important, in my view, is for the hocks to be reasonably large and well defined, and to be an identical pair. If on examination there seems to be an irregular feel to the joint, see if it is present on the other leg. If it is, and both the joints are cool and the horse is not lame, then all is probably well.

Other points of conformation, less likely to give rise to soundness problems, are nevertheless worth paying attention to when looking at a prospective purchase. The wither should be well defined, with a gently falling slope on either side. It should not fall away too steeply to the ribs each side, as it needs to provide a good bearing surface for the saddle to rest upon in order to prevent the saddle sinking down under the weight of the rider.

Many ponies have rather flat withers, as do some of the more common-bred horses. When a flat wither is combined with upright shoulders, the length of the stride will be restricted. Ideally, the angle of the slope from the point of the wither to the point of the shoulder is considered to be about 45 degrees in the riding horse. However, the length of the humerus bone (between the point of the shoulder and the elbow) has also to be considered, as a good sloping shoulder connected to an over-long humerus, which places the elbow further towards the rear of the animal, will result in a shorter stride.

Parents, when considering this part of a pony for their child, will do better to view it for correct saddle fitting rather than length of stride it allows. If the wither is flat and accompanied by a width of back that is very broad in relation to the overall size of the animal, it will give rise to problems – not only that of finding a saddle to fit, but also in getting the legs of the child to fit round the pony!

With larger horses the back should also come under scrutiny. A 'nice short back' is often ascribed to a good sort of animal; as the rider's weight is carried directly on the back it is reasonable to conclude that the shorter the back is, within reason, the better. This is because the shorter the back, the stronger it is likely to be. More important than the actual length, however, is that the horse should be 'well ribbed up' whether the back is short or relatively long. By well ribbed up I mean that the space between the last rib and the point of the hip is not excessive and does not reveal a large hollow gap. When it does the horse is said to be 'slack of rib'; backs of this conformation tend to be weak and prone to develop into a 'sway back' at a much earlier age than a more closely coupled one. However, as long as the back is strongly formed a bit of length is considered desirable, as it will provide a better, less jerky ride because the rider will be a little further away from the source of impulsion (the hindquarters). The horse will also cover more ground when striding out.

The neck should be straight, and in the riding horse not too heavy as it provides the horse's balance when moving. A neck that has a concave or hollow shape along its top edge when viewed from the side is referred to as a 'ewe neck'. Apart from looking unattractive, a ewe neck, especially in a mare, is often a sign of suspect temperament. The lack of muscle, or muscle that has been wrongly developed through bad training, will result in a poor head carriage and consequently a less pleasant ride.

The head, and particularly the eyes, should create a friendly impression and you should get the feeling that you and the horse will get on together well. The eyes should be large and clear, and be well set to give good forward vision. I like to see a bright, kind eye, as this usually indicates a good disposition. Small piggy eyes are associated with meanness, and animals that roll their eyes, showing the whites, are best avoided as they are likely to be bad tempered and unreliable. I also like to see alert-looking, pricked ears proportionate to the head in size. Droopy ears suggest sluggishness and long ears are considered by many people to indicate a horse with speed. If the horse has a pronounced 'bump' between the eyes, it is regarded as likely to possess an obstinate streak.

When examining the foot, we come to a very crucial part of conformation. The saying 'a horse is only as good as his feet' is a true one. No matter how nearly perfect all the other points are, nothing can compensate for faulty feet. In order to give this important topic the thorough coverage it deserves, I deal with it in Chapter 8, together with shoeing and related problems.

What type to buy

'All horses are nine' is an old joke among horse dealers, as it is the age that most horses are at their best. If it has been well cared for, a horse that is somewhat older than this can often be passed off as nine, hence this old saying. When trying to ascertain the correct age of a horse, it is well to remember that its outward appearance is not a good guide, as many well looked after horses will keep a young-looking appearance even when old. The reverse is also true: a neglected animal will appear to be older than his true age. It is therefore necessary to resort to some other method to determine the age of a horse, and this is done by examination of the teeth. To judge age correctly in this way is, however, no easy matter, and one needs to be pretty expert in knowing what to look for and interpreting what is seen. For this reason the true age of any likely purchase will best be determined by a vet.

When an advertisement gives an age as 'rising'—for example, rising six or rising seven—it means that the animal is nearer to the age stated than the year before it. Sometimes the word 'off' is used—for example, six off, which means that the horse is over six years old but not yet rising seven. Most horses and ponies are considered mature and ready for work at four or five years old, though some breeds—and some individuals—tend to mature later than this. The (usual) useful working age for most horses is from five to approximately seventeen years old. They are at their best at between about eight and thirteen. Having said that, many people will proceed to

A typical example of the good, all-round pony found throughout Great Britain.

argue that their horse went on for longer than this at full strength. I would not dispute this in certain cases, but for the purpose of anyone wanting to calculate what to buy, my general figures will prove reliable enough.

Unless you or your child is an advanced rider, able and willing to school a horse on, then one that is younger than six or thereabouts is likely to be insufficiently trained for you to enjoy, while an aged animal of more than thirteen will most probably not be able to give you the years of useful enjoyment that you will want from your purchase. I would therefore suggest something within the six to twelve age group as most generally suitable.

After deciding upon age, the next thing to consider is size. There is obviously an enormous difference between a child's pony on the one hand, and a horse capable of carrying a 14-stone (89 kg) adult on the other. As I have already explained, the type of animal as well as its actual height is an important factor in establishing its weight-carrying ability, short-legged types of a cobby build being able to carry more than the longer legged, more finely bred animals.

Without actually seeing an individual horse or pony and taking into account all the factors involved, it is very difficult to give a strictly accurate assessment of its weight-carrying capacity. Its height, weight, conformation, age and fitness must all be considered in order to determine what amount of weight the animal is up to carrying. However, for those readers with no practical experience of this who want some idea of how to judge whether an animal would be suitable in this respect, I have devised the following method of assessment, set out in detail in the table (Fig. 2).

Take a factor of between one-fifth and one-sixth of the animal's bodyweight and use this as being the approximate weight of rider and tack that the horse or pony can comfortably carry. Bear in mind that, as with humans, the smaller, stoutly built types will be stronger pound for pound than their larger counterparts. Begin by using one-fifth as the fraction at the small end of the scale, working down to one-sixth as you move up the heights. As an example, let us take a child's pony weighing about 600 lb (272 kg) and standing 13.2 h.h. One fifth of this bodyweight gives us 120 lb (54.4 kg) as a weight-carrying capacity. This would provide for a saddle, complete with stirrups and girth, weighing 12 lb (5.4 kg), and a young rider of up to 108 lb, or 7 st 10 lb (49 kg). Many strong breeds of pony will cope with more than this, and strictly accurate figures will also depend a lot on whether the pony is expected to carry the rider during a full

HORSE OR PONY			WEIGHT-CARRYING CAPACITY		
Size h.h.	Approximate weight lb (kg)	Divided by	Gross lb (kg)	Less weight of saddle lb (kg)	Weight of rider to be carried lb (st lb / kg)
13	500 (226)	1/5	100 (45)	12 (5)	88 (6 st 4 lb / 40)
13.2	600 (272)	1/5	120 (54)	12 (5)	108 (7 st 10 lb / 49)
14	700 (317)	1/5	140 (63)	12 (5)	128 (9 st 2 lb / 58)
14.2	800 (362)	1/5	160 (72)	14 (6)	146 (10 st 6 lb / 66)
15 (light horse)	950 (430)	2/11	173 (78)	14 (6)	159 (11 st 5 lb / 72)
15 (heavier type)	1050 (475)	2/11	191 (86)	16 (7)	175 (12 st 7 lb / 79)
15.2 (light horse)	1000 (453)	2/11	182 (82)	16 (7)	166 (11 st 12 lb / 75)
15.2 (heavier type)	1150 (521)	2/11	209 (95)	18 (8)	191 (13 st 9 lb / 87)
16 (light horse)	1050 (475)	2/11	191 (86)	20 (9)	171 (12 st 3 lb / 77)
16 plus (heavier type)	1350 (612)	1/6	225 (102)	22 (10)	203 (14 st 7 lb / 92)

Fig. 2 The figures given in this table are meant only as a guide and must not be taken too literally; much will depend on the conformation, fitness and constitution of an individual animal and the work it will be required to do in relation to the weight being carried.

day's hunting or on a long-distance ride, or merely spend an hour out on a quiet hack.

At the other end of the scale are the larger horses, and here let us take as an example a 15.2 hand horse weighing between 1000 and 1150 lb (453 and 521.5 kg). Using a factor of two-elevenths of bodyweight this would provide for a weight-carrying capacity of between 182 and 209 lb (82.5 and 94.7 kg). So, after allowing for the saddle, etc., a rider of between 166 lb or 11 st 12 lb (75 kg) and 191 lb or 13 st 9 lb (88.6 kg) might be carried. A 16.2 h.h. animal of 1350 lb (612 kg) bodyweight or more would be able to cope with a rider of 14 st (89 kg) and over, using a factor in this case of one-sixth of the horse's bodyweight. The same criterion of build and bone and of the type of work to be done also have to be borne in mind.

For those parents and adult riders who have not previously had to consider what size of animal was suitable, the foregoing will give an idea of the requirement and the chart I have devised should prove helpful. Parents should remember, though, that little Willie or Mary Jo will be growing and putting on weight and a pony of exactly the right size when bought will soon be outgrown.

The saddle weights given are only approximate and refer to English tack. Some saddles will weigh considerably more—the Western saddle and the Spanish Vaquero saddle, for example. I have been surprised at the weight of some Spanish saddles when lifting them off the rack, and would judge them to be at least twice the weight of their English counterparts. The heavy tack their horses are expected to carry is often, but not always, compensated for by the smallness of many Spanish riders, though many Western riders are very large men. Many of their horses are under 16 hands, which shows the importance of conformation and fitness when arriving at the weight a given animal can carry.

Who to buy from

Age and size having been decided, where does one find a suitable animal? Many horse magazines carry advertisements from people with a horse or pony for sale, and there are no doubt very many successful purchases completed in this way. There are one or two things to be wary of, however, and the first thing to consider here is distance. If the animal turns out to have a serious fault that was not mentioned at the time of purchase, then the more distant you are from the seller the more difficult it will be to obtain recompense. This disadvantage particularly applies to those advertisers whose reason for sale is 'Owner going abroad'. Of course, not everyone offering a horse or pony for sale who is shortly going abroad is trying to dispose of an unsound or unsatisfactory animal, but it is a point worth considering, especially if the horse is being offered at a bargain price.

When it comes to buying from dealers, I think it wise to try to find someone else who has bought from that particular dealer and ascertain whether they had a satisfactory deal. Many dealers build up a good reputation over the years, and they will often take an animal back without fuss if it proves unsatisfactory. This is not to say that if the horse goes wrong some time after purchase, perhaps through your own fault, that the seller will accept responsibility. Some dealers, and for that matter many private owners, are willing to let the animal be taken for a trial period; this can be a good arrangement. Part payment should probably be made in advance to show your good faith in wanting the horse if it proves suitable. Do get it clearly understood *before* you part with your money and take the horse exactly what arrangement is being made, so that no argument will arise afterwards owing to a misunderstanding.

It is said by many that buying from one's friends is best avoided if one values the friendship, and I think that for first-time owners horse sales are also best avoided. Quite often there does come a chance to purchase from someone giving up riding or seeking a horse with greater potential than the animal they have. If such a person lives fairly close and can be persuaded to let you try the horse for a week or so, this can often afford you the opportunity to obtain what you are looking for. Ponies that are outgrown are also often a good way to buy and it can pay to make enquiries locally to see what is likely to be for sale. Do not be afraid to ask questions, as most horse owners will give a candid answer and if they are not truthful then you have grounds for genuine complaint if things go wrong later. You will probably have a vet's opinion before buying, and this should help eliminate any major troubles. Some vets can be over-cautious in their opinions and although finding nothing to indicate that the animal is unsound will nevertheless notice something that they feel may lead to trouble later on. In this case the decision to overlook any minor faults may or may not be a reasonable risk, depending on how much you are paying for the animal and how hard it will be expected to work.

When it comes to methods of payment a lot of sellers prefer to have cash. There is nothing sinister in this; horse trading for cash has been practised for generations. If you pay in this way it is advisable to get a receipt with a fair description of the horse or pony, particularly its age, size and colour and any distinguishing marks, and it is also important to state that the purchase is being bought as sound if this is the case. A written warranty is the best form of safeguard, and this should ideally include mention of freedom from vice as well as soundness and suitability for the purpose sold. Ask the vendor to let you have a letter of warranty covering everything that is important to you in making the purchase. Age, soundness in wind, limb and sight, freedom from vice and suitability for the purpose for which you intend to use the horse or pony should all be listed. If the vendor is unable or unwilling to do this in full or makes qualifications over certain points, it is for you to decide whether this is fair and reasonable or whether he is trying to pull the wool (or in this case the horse hair) over your eyes.

It may well be that complete soundness is not as important to you as other things such as freedom from vice (bucking, for example) in a child's pony. This should be covered in the warranty by suitability of use, so make a point of discussing it at the time. The formula often used in advertisements that says a pony 'can be ridden by a girl' means nothing and cannot be taken to warrant the pony fit for this purpose. Likewise, 'believed sound' is no guarantee: it only expresses an opinion, and cannot be taken as a fact of warranty.

The actual wording of a warranty is important, and what is being warranted should be written following the word 'warranted'. To make this clear I will give you an example:

> Received from Mr Smith the sum of £750 (seven hundred & fifty pounds) for a seven years old bay gelding 14 h.h. and warranted sound, free from vice and quiet to ride.
>
> Signed ....

It will be seen that only those items following the word 'warranted' are included in the actual warranty and those that precede it are not covered. Thus the age and height in this case do not come within the warranty. Another point worth noting is that obvious defects in the animal are not covered in a warranty unless they have been discussed during the examination. Thus it is no use going back to the seller and complaining of a conformation defect that you only notice after you have bought the animal if this has not affected its soundness. Soundness in this case is taken to mean sound for the purpose for which the animal was bought.

The buyer will usually be expected to meet the cost of the veterinary certificate and any transport charges, though free delivery can often be arranged with the vendor if you try and then keep the horse. You should not, however, expect to be given a free trial and then to be given free transport both ways into the bargain.

Actual terms of payment, such as a deposit at the time of the deal and the balance on delivery or at the end of the trial, must be arranged with the seller. If you are expected to pay in full before delivery or a trial period and you agree to this, it is usually considered wise to pay by cheque. This is often the most convenient way for you to make payment in any case, but to do so as a means of safeguarding yourself so that you can stop the cheque if anything goes wrong is in my opinion of little value. Anyone who is unscrupulous enough not to be willing to put things right when selling a bad animal will have cashed the cheque before you have discovered your bad luck.

All that I have said represents the best way to make a purchase in theory, and if you can get fairly close to it in practice then you should not go far wrong. Do pay attention to the individual that you are dealing with and try to sum up his or her character. I feel that this is of such importance with horse buying that I would go so far as to recommend that you turned down what seemed a good horse if you took a strong dislike to the person selling it. On the other hand it would be foolish to go along to buy a pony costing four or five hundred pounds or several hundred dollars and expect that for that money you were entitled to a gold-plated investment with every possible safeguard. And don't assume that the vendor is out to cheat you, as outright sharp practice is not the general way of things with most horse trading through either dealers or private sellers.

Calculating the costs

The last thing to consider in this chapter is what your horse or pony is going to cost after you have bought it. The cost of its keep includes feed and bedding, shoeing, vet's bills, etc., all of which will steadily increase with inflation and cannot be avoided once you are an owner. There will also, of course, be the initial cost of tack and other equipment, which can add up to a considerable sum. I shall deal with ways to keep this to a minimum later in the book.

Feed and bedding costs will depend to some extent on whether the animal is to be kept stabled for most of the time or whether you are lucky enough to own or have the use of a paddock. Also, the amount of foodstuff that needs to be bought will obviously increase with the size of the animal. To give a fair idea of the actual overall cost involved, let us assume that we are dealing with a horse of about 15 hands that is to be kept stabled. You will need to allow a weekly average of 1 cwt (about 50 kg) hay and 56 lb (about 25 kg) corn or nuts. (This is not to be taken as the actual feed requirement for the horse, as correct feeding depends on what work the horse is doing and several other factors, all of which I shall be dealing with later in the book, but it provides a general indication of the average quantity consumed.) Find out the price of hay from a local farmer or merchant and also the price of corn* and cubes (cubes and nuts are manufactured horse foods of balanced ingredients), and you should have a good idea of the weekly food costs for such an animal. If the horse is much bigger than this, allow up to 50 per cent more; if it is a pony, then 25 per cent less. Having arrived at the figure to suit your particular animal, add another 25 per cent of it to cover veterinary bills and insurance and a further 25 per cent for food supplements, medicines and tack repairs. Another 25 per cent of the original figure will cover the shoeing requirements, and though the cost will vary depending on the type of bedding material used, a final 25 per cent will probably cover this item.

So, after you have worked out the cost of your hay and corn or cubes requirement, double this figure and you will have a fairly good idea of the total cost of keeping your horse or pony. If grass keep is to be used, then make the necessary reductions for the amount of time the animal will spend out at grass. If it is your own field, however, there will be the cost of its management and fencing, etc., and if it is rented there will be that cost. This can bring you back fairly close to the original doubling of the hay and corn costs.

Lastly, I am of course assuming that you will be carrying out all the work and odd jobs yourself. If you intend to keep your horse at livery, the type of livery and the charges made will have to be gone into; it is much more expensive.

*The word 'corn' used above refers to feeding grains such as barley and oats etc.

2 Stabling facilities

The ideal arrangement is to provide your horse with both stabling and a paddock and for these to be adjoining or reasonably close to your home. Many people fall short of this ideal and yet manage to get along quite comfortably without serious problems for themselves or their horse or pony. If you do not own any land yourself you will need to search around for something suitable that you can buy or lease. The availability and cost of suitable land will depend on the area in which you live. Perhaps you could find some old buildings on a piece of grassland nearby that could be hired and adapted or some unused stables that the owner is willing to rent out. There are many types of stabling that can be used, but they should all do the same thing: provide the horse or pony with weatherproof, well-ventilated accommodation that is easily managed. I shall concentrate on three alternative methods of achieving this: building your own stabling from scratch, using a variety of materials; converting existing buildings; buying and erecting ready-made sectional units offered by firms specialising in stables and loose boxes.

Before you begin it is essential to find out from the appropriate authority what regulations or restrictions there are to making provision for keeping a horse in your locality. (Things will be much easier in this respect for those living in the country than for those in urban areas.) Planning permission may have to be obtained and building regulations adhered to; these are best gone into before you start, or you may one day be told by an official that your newly erected stabling must all be pulled down. It is also wise to bear in mind the possible nuisance to others that may arise from anything you intend to do, and to try not to give cause for complaint. Most people, including neighbours and local officials, are open to reason.

If you adopt the same attitude, most things will go smoothly and quite often a satisfactory solution can be reached that will enable you to carry out your plans. Any objections that you do get to keeping horses will probably be from one or more of the following: nuisance from flies, smell and flies from manure heaps, or horse droppings on paths and roads.

Apart from causing annoyance to other people you do not want to suffer these inconveniences yourself, so think out the stable siting very carefully before you proceed. I have found that by spending a week going over a proposed site several times, beneficial changes to the original plan suggest themselves. Siting can often be changed so that the horse takes a different direction when leaving the stable, and dung-carrying arranged so that heaps are established where they cause the minimum of inconvenience. Hay and straw, too, will have to be delivered, and you do not want this done where it will blow all over the place close to your home or to that of your neighbour.

It is also worth checking whether the deeds to your property contain any restrictive covenant that prevents the keeping of livestock or animals of certain kinds. This does crop up occasionally, and if it applies in your case legal advice will be needed to see whether it can be overcome. If you rent or lease the property, your landlord will have to be consulted over this and the terms of the lease checked; permission from the landlord will anyway need to be obtained before you can go ahead.

Building your own stables

If you are a handyman, possessing the tools and the ability to construct a weatherproof building, your stabling will of course be cheaper to erect; it will also enable you to suit it to your personal requirements and to make the best use of the available space.

You will first need to consider what size of stable or loose-box is necessary. An internal measurement of 10 ft by 9 ft (3 m by 2.7 m) is about the minimum practical size for a small pony; 14 ft (4.2 m) square will be big enough for the largest horse. (10 ft by 12 ft (3 m by 3.6 m) or 12 ft (3.6 m) square are the popular sizes with custom-built units.) Go for a size larger rather than smaller than you think you need if you have the space, as this will make for better and easier management.

I like the loose-box to be big enough for the horse or pony to get down to rest or have a roll. If he is unduly restricted it will dis-

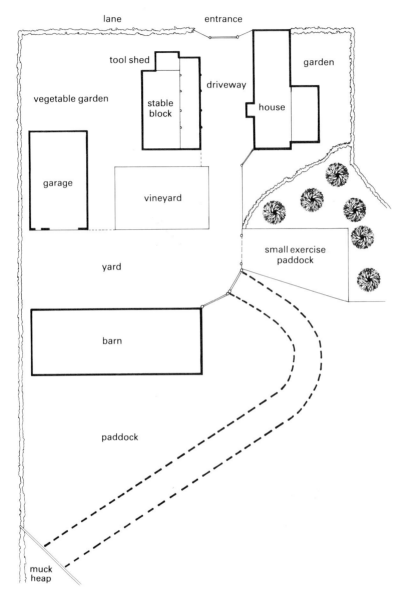

Fig. 3 Plan of the author's house, stables, etc. Note the position of the muck heap, situated some distance away from the buildings.

courage him from doing this, and if he does get down he is then more likely to get 'cast' or to damage the stable walls and fittings when rolling or trying to get up.

If your stables are to be erected at your home, then their position in relation to the house must be the right one. Having them fairly close to the house will not present problems if your construction and management are what they should be. The stable should preferably be sited down wind from your windows, and so that the prevailing wind and rain do not blow through the front door of the box. My own stables are just across the drive from the house, opposite the main windows, and this has not raised any problems regarding smell or flies. Easy access to the stable is necessary for both feeding and mucking out, and a good hard surface surrounding the area is essential to avoid working in mud. Because I have built or converted my own stable buildings, and they work well, I have included illustrations in the plate section that will show my arrangement. They are not perfect, nor will everyone have the space to plan theirs likewise or the need for such extensive stabling. But they do provide a model for anyone planning his own, even if this is to be on a smaller scale.

When considering your site, don't be put off if you do not have enough level ground available. One man with an earth-moving machine can work wonders in just a day! However, if you need to do this, consider the surface water drainage and if in doubt consult an expert before you have any large area dug out.

Suitable building materials

The type of materials that can be successfully employed range from concrete blocks or bricks to all-wood or wood framing with corrugated sheeting as cladding. Concrete blocks make good walls, those of 6-inch (15 cm) thickness being better able to stand up to horses than the 4-inch (10 cm) ones. They are cheaper and easier to construct than bricks, though they need to be well soaked before use to ensure that the mortar 'sticks'. A good foundation must be provided for walls built with blocks or bricks; when foundations are being dug out, do not overlook the need for drains, as provision for them at this stage will be easier than providing it later on.

All-wood constructions are fairly easy to erect. One good method is to build a framework of 3-inch by 2-inch (7.5 by 5 cm) timbers and then to nail 'shiplap' planking to the outside. This type of planking interlocks, and each plank has a concave lower edge to carry off rainwater. For the interior, a very good cladding is obtained by using large sheets of special density chipboard of the type used for the upper floors of housing. Before this is nailed onto the interior of the framework, make sure that you work out where the stable fittings are to be fixed and put some extra pieces of timber behind the chipboard for this purpose. Take the measurements and draw a sketch plan so that you know where these are after you have fixed the chipboard in position!

Stable doors can be similarly constructed, but it is best to clad these both inside and out with the shiplap, as they often stand open and get wet and chipboard does not react kindly to this. Paint all the woodwork with creosote or tar varnish to preserve it. This is messy, but still the cheapest and best treatment for wood. It has the added advantage of (sometimes) deterring horses from chewing at the woodwork because of the nasty taste.

It may be cheaper to use corrugated steel for the outside walls, especially if good second-hand sheets can be obtained. I would not advise the use of asbestos sheeting for this; asbestos is very brittle and difficult to work with when it comes to sawing or drilling and nailing. Asbestos is also not a good material for roofing, because of both this brittleness and its poor weight-carrying ability. This is an important consideration, as even during mild winters there can be periods of snow and a good layer of wet snow is very heavy, easily causing a weak roof to collapse. If you do use asbestos, pitch the elevation of the roof as steeply as possible to offset this problem. Roofing felt over wood (solid planking, not open framework) makes a weatherproof and long-lasting roof. Clay or cement tiles on a good wooden roof fixed onto battens is even better, and also provide the best insulation and warmth.

Corrugated mild steel also makes a good roofing material, and can be easily fixed onto 4-inch by 2½-inch (10 by 5 cm) timbers set 2 ft 6 in (75 cm) apart. Arrange the timbers for the roof so that the corrugated sheeting is supported by them across their width. Tar both sides of the corrugated sheeting to prevent condensation and rusting; use galvanised nails fitted with plastic 'cups' to prevent rain from entering the nail holes. Nail the sheets through the raised parts of the corrugations, *not* in the valleys. Overlap each sheet by two corrugations, and fit these so that the overlaps are down wind; this will prevent rain from being driven in underneath. Similarly, if more than one sheet is required in length down the slope of the roof, don't forget that the lower one should fit underneath the one above where they overlap.

If you use second-hand corrugated sheets you must build your wooden framework so that it matches up with the existing holes in the sheets. Consider this point when you inspect the material before buying it, as if there are too many holes in it or they are unduly large because the sheets have been stripped off carelessly your task will be at best difficult, and at worst you may find that the sheeting is useless when you try to erect the roof.

It is sometimes said that corrugated steel sheeting should not be used as a roofing material as the summer sun causes it to heat up and make the stable very hot and stuffy. I have not found this to be so. If the building has been properly thought out, with adequate ventilation, there should be no problems. A slight disadvantage is that condensation does occur under certain humid weather conditions, but a good application of tar to both sides before assembly helps to reduce this. The gaps left by the raised parts of the corrugated sheets at the ends will provide an outlet for hot air to escape when summer conditions make the roof hot. It will also help to remove unwanted fumes and help to keep deep-litter bedding dry and well aired.

My own main stable block has a pointed roof of interlocking wood, battened across, with clay pantiles on top. This is excellent both for insulation and to look at, though it is rather costly compared with other alternatives. It also has louvred ventilators built along the top: any stale air or ammonia fumes that would otherwise collect at the top can easily escape.

Flooring

When it comes to flooring, one need not look beyond concrete. It is very hard-wearing, easy to lay, and reasonably cheap. If you mix it yourself, add enough cement to ensure that it will be strong enough to stand up to abuse; if you have ready-mixed concrete delivered to the site, make sure to order the 4:2:1 mixture or whatever is recommended for hard use. When laying directly on top of a dirt surface, see to it that the dirt is hard-packed. To strengthen and support the finished floor, get some old steel angle or metal bed frames and lay these across the area before putting down the concrete. Alternatively, put down several inches of hardcore to form the base, knocking it well down before applying the concrete. Whichever method is used, hose the base with water before laying the concrete to ensure that the two marry together. Failure to do this will mean that the floor will become 'drummy' in time.

You should allow approximately 4 inches (10 cm) of concrete for the thickness – less than this will crack up. It is imperative to lay the floor absolutely without hollows and to provide a continuous fall of about 2 inches (5 cm) to 12 ft (3.6 m). Don't put the concrete down too wet as this will weaken it. Use a length of 4- by 2-inch (10 by 5 cm) timber to tamp it level, and check with a spirit level as you go that you are maintaining the correct amount of fall. Before each area gets out of arm's reach, take a cane and scratch lines to form 4-inch (10 cm) squares in the wet surface of the concrete, making them a ¼ to ½ an inch (6 to 12 mm) deep. This will assist drainage and also give a good grip to the finished surface. Allow at least several days for the concrete to dry out before using it; if you do not, it will still be 'green' and the top will scuff up and leave hollows.

Surface drainage should be continued outside the box and incorporated into a concrete apron. Make the apron as large as possible as this area will be in constant use, and the more hard standing one has the better. The fall of the stable floor should slope towards the front door of the box, as should that of the apron. When laying the apron, make a shallow gully about 6 inches (15 cm) wide along the outside of the box: this will collect the water from inside and out. Make the gully so that it falls away to where it can run into a drain, and pipe the drain away underground to a soakaway large enough to take a week's rain without swamping and overflowing back up the drain – a hole 3 ft wide and 3 ft deep (say a metre each way) will be the minimum required, and it should be filled with old brick rubble. The stable guttering should be fixed so that the downpipe also empties into the gully: this will help to keep the drains 'sweet'.

You will probably have seen stables where the drainage is underground, with the floor falling towards a drain in the centre of the loose-box. This is not a good idea: it is smelly, easily blocked up, and difficult to clear.

Ventilation, door and window heights

Good ventilation to expel unwanted fumes and stale air is of the utmost importance. A stable should provide protection from the elements but it should not have a muggy atmosphere; equally, it must not be draughty. If the loose-box is built with a pointed roof a louvred ventilator can be put in the top, although this is rather tricky to build and keep waterproof. An easier and equally effective method is to provide some louvred slots high up in the gabled end of the box. These can be made from weatherboarding.

A window should be provided if the horse is not to be condemned to a black hole of Calcutta. If this is not to open, a thick reinforced plastic that has a wire mesh inside it can easily be arranged to provide the window. If the interior of the box is not particularly high it may, however, be advisable to have a window that opens to give extra ventilation. Special stable windows can be purchased with a top section that opens inwards by dropping down into metal guides that leave it tilted open. These need to be adjusted according to the weather, as in certain conditions the open window will cause a blast of cold air to enter the box. Side draught from this opening is particularly unwelcome.

Do be sure that your stable is high enough to allow the horse to stretch his head up without striking cross-beams or the actual roof. A height of about 7 ft 6 in (2.3 m) to the top of the side walls, with a further 4 ft (1.2 m) up to the point of the roof, will generally be about right. Windows should be positioned so that any opening is above the height of the horse or pony's back, as any draught that is produced will then cause less discomfort. If the window is fitted with glass or with the reinforced plastic described, iron bars no more than 6 inches (15 cm) apart should be screwed in front; better still, fit a thick wire mesh, fixed clear of the window. This will prevent the horse from poking out the glass or pushing his head through the plastic and scratching his face on the wire reinforcing. Many horses like to press their noses against these surfaces, and a costly repair will eventually result if not injury to the horse as well. The actual size of the window will depend on what fits in with your particular construction and/or what you can acquire cheaply; a good average size is 3 ft 6 in deep by 4 ft wide (1.06 m by 1.2 m).

The size of the door required can also be scaled to suit your needs if you are building your own loose-box. The minimum width required for a stable door is often said to be 4 ft 6 in (1.37 m). The reason given is that this width prevents the horse from striking a hip when being led in or out. I find this rather excessive, and although one needs to avoid a narrow entrance my own doors are only 3 ft 2 in (just under 1 m) wide and have never caused such problems. The wider the door, the more difficult it is to support the weight and avoid it dropping its leading edge and scraping the ground. This can be most annoying; I like doors to open and close easily, particularly as one often needs to get in or out of the stable quickly. We are all familiar with the saying that it is no use closing the stable door after the horse has gone: a dragging door may be just the thing that prevents it being closed in time. A width of 4 ft (1.2 m) is the maximum width necessary, and this will be adequate for the largest horse.

The height of the lower half of the door can also vary according to the size of the occupant and the type of bedding used. It needs to be high enough to prevent the horse or pony from attempting to jump out. Mine are 4 ft 4 in (1.3 m) high, and to make sure that there is no temptation to jump over them I suspend a safety chain across the top opening when the horse is on deep litter. (On this type of bedding the floor inside the box is raised by 6 inches (15 cm). I have also incorporated into my main block of loose-boxes doors that open out from the rear of each box in addition to the front doors. I added them afterwards, but anyone intending to adopt this practice would best provide them when first building the box. (The reason for including them is explained in chapter 3.)

A stable built of concrete blocks 6 inches (15 cm) thick, the best size for stable walls and partitions. If the hay-rack is correctly fitted there should be no problem of seeds getting into the horse's eyes—indeed, the animal's neck muscles will benefit from being stretched and exercised. The wooden board screwed to the wall behind the manger prevents the horse from lifting the fibre-glass manger out of its metal hanger.

Stable fittings and fixtures

The next thing to consider is where to place the interior fixtures that are necessary. These should be kept to an absolute minimum, as the fewer obstructions to free movement that a stabled horse has the better. I like the fixtures to be fitted in the corners as they are less easily walked into by a 'swarming' animal. Fit them high enough to be out of harm's way from the flying legs of a horse that is rolling and so that he cannot bang his head when lifting it up after picking about on the floor. For horses of between 15 and 16 hands a good height for an automatic drinker, for example, is 3 ft 9 in (1.15 m) from the floor to the top edge; ponies will of course require it lower. If a water bucket holder is to be fixed to the wall instead of an automatic drinker, the height given will prevent a rolling horse from kicking the bucket out and getting its leg stuck through the metal holder. It is not a good idea to leave a water bucket on the floor as its contents are so easily spilled. The manger should be fitted a little higher because of the extra depth of this compared to the automatic drinker. A hay-rack, if fitted, should be about another foot above this, at 4 ft 10 in (1.5 m) to its lower part if it is the corner type: the rectangular types should go lower alongside the manger. Many books advise the reader against the use of hay racks because of seeds getting into the horse's eyes; I have never had any trouble from this, however.

Deep litter bedding will build up in height and the standing horse will be that much higher. Allowance will have to be made for this, 2 inches (5 cm) or so being added to the heights given. As the litter will not be at its highest where the animal's legs are standing, this will be enough and it will also not place the fittings too high before the deep litter builds up again after being renewed.

Another requirement is for a strong tie-ring to be fitted into the wall so that your horse can be tied up for grooming, etc. I provide two, one on either side of the box, so that I can position the horse where I can easily get in and out; these should be about 5 ft 6 in (1.7 m) off the floor. They are also used to tie up hay-nets when these are used.

Provision of electric light is an absolute must for both the loose-box and the feed room. Constant use will be made of it on dark winter days and for the emergencies at night that do occasionally crop up no matter how careful you are. In the loose-box you must arrange things so that the horse or pony cannot get at the cables and light bulbs to chew on or play with. Light switches are best kept

A loose-box with fittings correctly positioned and showing hard standing for day use. The concrete floor offers a good grip; walls are lined with high-density chipboard above a 6-inch (15 cm) skirting board that is less affected by moisture. Note that the water pipe to the drinker is lagged and boxed in for protection.

A tie-ring at the correct height for tethering the horse and for tying up a hay-net.

outside the box, well beyond the reach of outstretched necks. Cables should be fixed as high as possible and concealed behind wooden battens and light bulbs situated out of reach or given a stiff wire cage for protection. Make sure that there is a power point in the feed room so that a kettle or a wander lead can be plugged in.

Converting existing buildings

The prospect of doing a conversion always excites me. There is much that can be done to make existing buildings suitable for horses. If you find the prospect rather daunting, do not be put off if what exists is badly run down; within reason, the older and more ramshackle the buildings are the better, as it will leave plenty of scope for improvement and alterations.

First and foremost inspect the main superstructure, paying particular attention to the roof supports. Make sure that the uprights are sound and possess sufficient strength to take any additional weight that they will be required to carry. Weak uprights can often be strengthened by fixing lengths of steel angle to them. The metal frames of dismantled army beds can be used for this, and can often be picked up cheaply. Any uprights that are too far gone will have to be replaced; when this is necessary do it before anything else is attempted. Next, look to the roof itself– and don't go clambering about on top of it until you are certain that the beams and rafters are able to take your weight. If these are in bad condition, it is false economy to try to patch them up: remove what you can of the tiles or other cladding and fix new roof timbers.

Once a sound roof on good supports is achieved the rest of the work can go ahead whatever the weather. First, though, look to the drainage and plan the entry and exit of horses and vehicles to avoid mud. If the weather is dry when you do this, try to visualise what it will be like after a bout of continuous rain. I cannot stress this point too much; if this book does nothing more than to persuade you to follow this one piece of advice, you will be eternally grateful.

The 'American barn' type of building makes excellent housing for horses when converted. Most barns have high roofs, and looseboxes constructed inside them will have plenty of air without draughts. The existing flooring found in these barns ranges from dirt to neatly laid cobblestones. Depending on what is there, you may find it necessary to renew only part of it; in any event the principles behind the construction of new stabling should be followed as far as possible. Very often it is also possible to provide for hay and straw storage, as well as feed and tack rooms, all under the one roof. This makes an ideal arrangement. If in addition there is a small piece of land or a paddock adjoining, consider using some of this to provide a small exercise area for the horse or pony. This can be arranged so that the animal can gain access to it direct from the box. When it is necessary to leave a horse or pony confined to his box for long periods, this additional small area that he can wander in and out of will help to prevent boredom.

Buying ready-made units

This will be the most expensive way of providing your horse or pony with stabling. Another drawback is that although manufacturers offer a wide choice, and the designs are usually very good, it is unlikely that one firm will offer every detail that you prefer. Individual requirements can sometimes be specified, but this usually increases the price considerably.

Having chosen the version you most favour, at a price you can afford, some saving can be made by putting down the concrete base yourself. Prepare this in the way already described, and surround the area with stout planking to contain the wet concrete to the required depth. Some sectional buildings need to be bolted down to the base as they are assembled; this holds the sides in position and fixes them down. If you are doing the preparatory work yourself, ask the supplier of the building to let you have a set of plans so that you know precisely where these bolts are to go. The concrete base can be drilled out after it has set, but it is much easier to push the bolts into the wet concrete as it is setting off. A word of warning, though: they must be *exactly* right in order to match up with the holes in the 'plates' of the building sections.

It is also worth incorporating a damp-course between the concrete floor and the wooden sides of the box. Rolls of special damp-proof material can be purchased for this, and are easily placed into position as each side is assembled.

Provision for hay and straw storage

Space for this will have to be found under cover, and again this area can hardly be made too big. Most people, however, will have to make do with something that is less than ideal. Good planning will often go some way to offset this. Depending on your particular circumstances, it might be worth considering adding an adjoining structure to your loose-box rather than erecting different types of building. Most feed merchants will not deliver less than five tons of hay, as it is uneconomic in these days of high transport costs for them to do so. Five tons occupies a considerable space, requiring a building about 16 ft (nearly 5 m) long by 12 ft (3.6 m) deep and

12 ft (3.6 m) high. A bale of hay weighs approximately half a hundredweight (25 kg) and measures about 48 by 20 by 16 in (120 by 50 by 40 cm). These figures will enable you to work out how much you will be able to store in your available space. Straw bales are lighter and smaller: 42 by 20 by 16 in (105 by 50 by 40 cm) is usual. Unless you have a lot of space, it would be best to find a local farmer to supply you with a smaller quantity at a time. Even he will not be interested in too small a load, so try to be able to take at least forty or fifty bales at a time. Never store your hay directly on the ground, even if this is on a concrete base under a good roof. In a very short time the bottom bales will develop musty mildew and be useless. Keep it up off the ground, preferably on a wooden platform with air circulating beneath. Straw, too, will suffer; not so badly, because of its coarser nature, but musty straw is equally useless. If it is used as bedding the musty spores will irritate the membranes of the horse's lungs and air passages, and constant subjection to this will inevitably lead to wind troubles. Poles laid across concrete blocks will provide a cheap and efficient method of storing straw bales, and hay bales can be stored on top. This will keep both in good condition and you can use them up section by section as you feed and bed down.

Water

You will need to provide water to the stables. This is best done by using the special agricultural fittings that are generally available for this purpose. A heavy-duty plastic tubing that is both flexible and lightweight can be used for the water pipes, and it is suitable for all connections either above or below the ground. It is easily cut with a hacksaw and all the joints and angles for any fitting are obtainable. Any competent assistant at the suppliers of agricultural equipment in your area should be able to advise you on what is needed and how to go about fitting it together. Run your water supply to each loose-box if you intend to have automatic drinkers, and provide a handy stand-pipe for filling buckets. Lag the pipes to protect them from the weather and box them in to protect them from your horse. Don't forget to include a stop-cock so that you can turn off the water supply to the stables in case of trouble. If this is overlooked and repairs or alterations are needed you may be faced with having all the water supply to your house cut off when any stable work is done.

Tack room and feed store, showing a portable saddle stand and a fitted storage rack made from scrap materials; note how the metal hanging-rails are slotted into the wooden uprights. The electric cooker is fitted neatly in the corner.

Tack and feed store

A good strong weatherproof room is essential for a tack and feed store. Ideally this should adjoin the loose-box so that everything is conveniently handy. Try to allow enough room for feed storage bins, a good-sized table, a saddle rack and a cupboard for cleaning materials, etc. You will also need space for wet riding macintoshes and muddy boots, as these are best not trailed and trampled about the house. Girths and saddle cloths will also have to be washed and hung up to dry, so the provision of a sink will be an advantage. This will also be needed for cleaning feed containers, etc. If hot water can be provided then so much the better. A good, strong, locking door and barred windows are desirable to protect your expensive tack. If you are in any doubt about this, take at least your saddle and bridle into your home at night.

3 Managing your stable

If you go about it in the right way your stable will keep clean, the work involved will be enjoyable, and it will reflect the pride that you take in owning your horse or pony. Gone about in the wrong way the task will be difficult, you will not enjoy doing it and things will become neglected. It is therefore important to choose the method and materials that best suit your particular circumstances.

Types of bedding

Three principal types of bedding material are widely used: wood shavings, peat and straw. Other types do exist, but unless you have a special reason for using them it is better to choose from the three I list.

Wood shavings

Provided that a regular supply of shavings can be obtained, and they are not wet when you get them because they were stored in the open and soaked by rain, they do make adequate bedding. The bed must be deep littered to at least 6 inches (15 cm), preferably deeper, in order to provide sufficiently resilient ground cover. There are, however, several drawbacks. Although I refer to shavings and not sawdust, even shavings will contain a considerable quantity of dust. This dust is aggravating to the eyes of both humans and horses and is harmful to the air passages and lungs. It is also said that the horse's hooves tend to heat up on both sawdust and shavings, though I have never met anyone using them who complained of this. A more serious problem, both with sawdust and with shavings, is that they frequently contain nails and even pieces of glass. Not only the feet of an animal but also the body would be at risk from these, from cuts and punctures inflicted when the horse is rolling. If shavings are used the droppings should be picked up and wet patches dug out regularly, and dry shavings added to make up the deficiency. The bed will need to be removed completely every three to six months depending on how well it has been managed. As it is useless for manure, and only makes a slimy mess if spread on grassland, it will have to be dumped and then burned when the weather is suitable. Another drawback is the way it spreads when it is trodden on, and forms slippery deposits on gravel drives and paths.

Peat

Many people advocate the use of peat, though I am not among them. However, it is widely used. It, too, must be deep littered, and must be completely dry when first put down. As it is usually supplied in large, sealed plastic bags keeping it dry is no problem. In fact, a considerable advantage to those possessing only a limited amount of cover is that these bags can be stored in the open. A disadvantage is that they are very heavy and the smooth plastic surface provides nowhere to grip when moving them. When the bags are wet and slippery this task is made very difficult indeed, and no woman or child could be expected to cope with it. About ten large bags will be needed to provide a deep-litter bed for the average stable, and approximately three bags per month will be required for topping up. Once the bedding is down its management is the same as that of wood shavings. Apart from nails and glass, which it does not contain, it has the same drawbacks: it is easily trod about, is dusty and can irritate the eyes. I have seen it claimed that peat acts as a deodorant and eliminates ammonia smells; I have not found this to be so. I have found that it absorbs and then retains moisture, and even when maintenance is good there is a degree of staining to the animal's coat. Because of its tendency to hold moisture, many examples of it in use that I have seen were quagmires even when drainage and management conditions were reasonable. I also regard it as expensive, though many people would disagree and claim that its initial cost can be offset by selling it as manure after use. I question the economics of this practice; it may make very good manure when mixed with horse dung, but the amount of time and money that must be spent in order to sell it for this purpose outweighs the return. It must be put into suitable plastic bags; there is the cost of advertising and then delivering it; and the possibility that when you arrive the old lady who wanted it for her garden is

The straw bedding is banked up around the sides of the box to prevent the horse from getting cast. The tidy threshold fitted across the entrance prevents the deep littered straw from scattering out onto the concrete apron. The photograph also shows the surface drain incorporated in the apron.

either out or expects you to carry it a considerable distance. If you manage to sell it for enough to cover your time and labour you will be lucky indeed. Without a profitable scheme to recoup costs I reckon peat to be twice as expensive for bedding than straw.

Straw

Provided that it is clean and free from mould or spores, straw is probably still the cheapest and best bedding material. However, it must be emphasised that any old straw will not do, as anything containing damp mould or spores will do great harm to the horse's lungs. Many horses that develop coughs do so because of no other reason than being bedded on bad straw. The most widely available types of straw are barley, oat and wheat. Barley straw is too soft and will become squelchy; it also tends to be eaten. Oat straw is better, but again will be eaten by most horses and ponies. Wheat straw is the best as it is the hardest and less of it will be consumed. Many old-time horsekeepers complain that straw today is so badly mutilated by modern machinery that its value for bedding is lost. No more do we get the long, hollow-stemmed reeds that can be layered to provide a resilient, well-drained bed. Instead it is short, very compressed and much less resilient. This is true, but in my view it is still the best bedding material especially when used for deep litter, when the loss of these qualities is much less important. (What concerns me more, especially if some of the straw is eaten, is the possibility of the animal ingesting the toxic pesticides that are widely used in modern agriculture.)

The deep litter bed

If it is to work well, deep litter straw bedding has to be as carefully managed as other types of bedding. Droppings must be removed when seen and not left to accumulate and be trodden in, and the wet parts of the middle of the bed should be taken out daily. Fresh straw will then have to be added. When starting a new deep litter straw bed you merely spread several bales of straw over the clean floor and allow the horse to compact this down. Add a bale of fresh straw every day after removing droppings until the bed builds up into a compressed mass several inches deep. After this only about half a bale of straw or less will be needed daily to make up for the loss of that removed. When removing wet and dirty patches be sure not to disturb the bed to such an extent that the deep litter base is broken up. Properly done you should finish up with a good compressed deep litter base with a top layer of clean loose straw.

I consider my own method of deep litter wheat straw using the separate day and night box system to be the best. But – and it is a big but – this can only be done if you can afford to have two loose-boxes for one horse. It works like this: one box is kept as deep litter, and the adjoining box merely has the hard standing of the concrete floor. The stabled horse is taken from his deep littered night box after breakfast (sometimes before), and uses the hard standing of the day box until changed back at late afternoon for the evening feed. Some people have told me that their horses will not stale if stood on a hard surface because they dislike splashing their legs. I have never had any problem with this and believe the reason some horses dislike staling when on hard surfaces is because they have slipped when spreading their legs. I find that having good grooves in the concrete floor gives them confidence and a good footing and consequently my horses stale regularly. After the night box is cleaned, it is left to 'air' and dry out during the day. The fresh straw replacing what was removed in the morning is added just before the horse returns to the box in order to let as much air as possible get to the bed. I have found that managed in this way the deep litter base can remain down for six months or more, and I have never been troubled by bad smells or ammonia fumes, to which my nose is especially attuned. (It was explained to me by a friend of mine who managed Thoroughbred studs for many years that a chemical reaction in the straw neutralises the ammonia, hence the absence of any problem. I am unaware of any scientific proof of this but it would certainly appear to be true.) My horses look forward to being changed over to their night boxes, and trot happily in and are soon down having a good roll. I believe that this has good psychological value and relaxes them ready to enjoy their evening meal.

I would unhesitatingly recommend this as the most effective and labour-saving method of keeping a stabled horse or pony to those with the facilities to employ it. Failing this, a combination of deep litter straw in the stable and turning the horse out during the day, or a single box deep litter or part of it deep littered system may be used. Another point in favour of deep litter is that it is warm and comfortable for the horse and encourages it to lie down. When it gets up again the bed provides a good footing and prevents the danger of the animal slipping or scraping itself on the floor. It is for these reasons that I prefer it to the more traditional use of straw management for bedding.

Other bedding methods

Some people prefer to use straw in a single box and muck this out completely every day. The floor of the box will thus be left uncovered to air and dry. Others muck out the worst in the morning and leave half the box uncovered; the bedding will then be relaid with fresh straw, perhaps including some of the best that was taken out and dried off, which will provide the horse or pony with his night bed. Probably the most traditional and widely used method is to muck out in the morning, leaving most of the good straw banked up around the sides ready to go down again and provide a thickish night bed; the centre area of the box is left with a light covering of straw for its daytime use. These methods also work well and there are many well-managed and properly run stables that use them. However, I must say that I have seen horses encouraged to get down by a thin layer of straw who then had to struggle to get up again because it was pushed away, leaving a bare floor, when this was attempted. Straw beds should also have the straw banked up around the sides of the box to prevent a rolling horse from becoming cast. This banking will need to be layered fairly firmly with the fork. If you use deep litter straw you will find that a natural slope of compressed straw will build up around the edges with correct management and this provides an excellent barrier.

It is also advisable to bank up the new straw beneath the automatic drinker or water bucket holder. Some horses find these nicely placed to back onto and rub against, and will sometimes lift their tails and deposit a dropping into the water. This makes a filthy mess and is an unpleasant cleaning job so prevention is highly desirable. Mangers, too, can suffer in this way, and I have found that banking up the straw beneath nearly always prevents this. I have seen objections to this practice, because it is claimed that worm larvae can climb up and get into the manger, where they will be eaten by the horse and recycled. I do not have the straw high enough for this, but in any case do not agree with this assumption–horses will inevitably pick about for food in their bedding, and any larvae that are present are more likely to be ingested this way.

The muck heap

Whatever system you decide upon, if using straw you will need to establish a muck heap. With only one or maybe two horses, this will not need to be very large and will continually rot down. However, these heaps do tend to spread and the surrounding area becomes messy and unsightly. For this reason it is best to have a pit dug out, which will help to contain the muck and assist in its rotting down. Burning the drier surface layer will also diminish it. Have the site as isolated as you can from the house but make sure that it is accessible even in winter. The rotted down part can be used for manure, and if you find that you have too much for your own use it can be offered to friends and neighbours willing to collect it.

Cleanliness

A spell of fine weather should be chosen as the time to remove a deep litter bed completely. The floor of the box should be swept clean and given a good scrubbing with disinfectant, and then be allowed to dry out completely before a fresh bed is laid. Day boxes and those where hard standing is used for the horse should have the droppings picked up when seen and be swilled out daily when the horse is absent. Once a week all the cobwebs should be brushed down, together with the dirt they collect. Some people say that these should be left as flycatchers. This is nonsense – the number of flies caught in this way will be very small. Remove and clean all mangers and water containers once a week; if they are not removable give them a thorough cleaning where they are.

Another good idea, essential for any form of deep litter, is to fit a 'tidy' threshold. This is a block of hardwood about 6 by 4 inches (15 by 10 cm) thick, fixed down across the doorway. It can either be fitted across the door uprights at ground level or, better still, bolted down to the concrete floor, where it will also provide a draughtproof door stop. Where they can be incorporated, rear doors in the loose boxes are another aid to good management. I have them leading out into a covered area behind the stables, from where I attend to all the daily routine. I can muck out under cover directly into the dung trailer or wheelbarrow, and enough hay and straw is kept here to last for ten days. Because this area is concreted the task of sweeping it up and keeping it clean is very easy.

Ventilation

A stabled horse or pony must always have fresh air available to him. When one considers that if the tissue from a horse's lungs could be spread out it would cover the area of a tennis court, the volume of air that needs to be continually changed will be appreciated. Good ventilation without draughts is very important and one should pay particular attention to window and top door arrangements, always bearing the weather in mind. I find that with both a front and rear top door to open I can regulate these according to the prevailing wind and so have one open at a time without getting a blast of cold air through the box. On hot summer days and nights both can remain open to keep the box cool, and on cold winter ones they can both be closed and the window and ventilators used to expel stale air. If the horse chooses always to stand in the same place on cold or windy nights he may be avoiding a draught of which you are unaware. If the animal stands resting quietly on his night bed, lays down occasionally and looks generally happy and not dejected, it can be assumed that he is comfortable and the bedding is adequate. On the other hand, if he looks miserable, the hair over his back and sides stands up on end (called staring) and he hesitates to get down and when he does he slips on trying to get up, you had better think again.

Blankets

Although blankets are not directly related to types of bedding, I include them in this chapter because it deals with the horse's warmth and well-being. Let me say at once that as a general rule I dislike blankets and rugs for a stabled horse. If my advice is followed regarding stabling, bedding and management there should be no need to use blankets for the normal horse or pony. They should only be necessary for an animal that is sick, fully clipped for hard work or possesses a very thin skin. Some people, believing that they are being kind to the animal, pile on a great number of rugs and blankets at the first sign of autumn. I have seen some horses positively sagging beneath their weight! Other owners will use them to keep the horse clean or to try to prevent it from growing a full winter coat. It is also claimed that blankets save food by conserving heat loss, thus saving the cost of extra food that would otherwise have to be given to provide extra energy for heat. I do not agree that this practice saves costs. One has the expense of the blankets, surcingles, etc., to which the cost of cleaning and repairs has to be added. Consider, too, the energy needed to lift the heavy weight of this clothing up and down every time the animal's ribs expand with breathing. I firmly believe that the great majority of horses would, given the choice, elect to be without this additional encumbrance, even if it meant the odd night or two being less warm than they would like. Don't misunderstand this: I am not advocating leaving a poor animal confined to a cold and draughty or leaky box, unable to keep warm. If you cannot provide an airy, draughtproof, comfortable loose-box, you will need to provide a blanket when the weather is bad. When you do use blankets make sure that they fit properly, do not rub – especially over the withers – and that they are not left to become the filthy, smelly things that are too often seen. For night use, when necessary, a rug with surcingle attached for keeping it in place can be obtained. Plain rugs with a separate padded surcingle to prevent pressure on the spine can also be bought. Another device for keeping a rug in position is an arched roller, which fits across the spine rather like the tree of a saddle, thus avoiding pressure on the spine and also preventing the horse from rolling over and becoming cast. Whichever method you decide to use, pay attention to the fit and only use rugs and blankets when you are sure that something more is needed even though you have done everything else to make your horse or pony comfortable.

Other stable tips

Some horses and ponies become adept at slipping off their head-collars, untying ropes and opening doors. The practice of opening the top door is particularly annoying. My top doors have a latch fastening that allows me to enter the box, bolt the lower door behind me and then also close the top door from the inside when required. I am thus able to open it again to let myself out. The disadvantage is that some horses quickly learn to operate these latches and can therefore open their top doors. Being awoken on a wet and windy night by a banging top door and having to dress to go and close it is no laughing matter, and was something I quickly decided to cure after the first instance. Obviously a bolt can be fitted to the outside, but as I already had a chain that is suspended above the bottom door when the top door is open, I merely adopted the practice of clipping this across the closed top door at night. This prevents the top door swinging open if the latch is undone, and I have found that since the animal cannot get his head out he soon does not bother to fiddle with the latch.

Another very useful trick that a horse will learn is to stale in a bucket that is held beneath it. When returning from exercise hold an empty bucket and give a series of short whistles to encourage the horse to stale. If this fails to produce the desired result, scratching the floor between its hind legs with your foot or rustling the straw will often bring about what is required. This routine will soon be learned if it is practised regularly. I have taught my horses to do it when they change over to their night boxes in the evening, and so

Top door latches that can be opened from the inside are useful; a chain clipped across the front of the closed door prevents it from swinging open if the horse manages to unlatch it, and when the top door needs to be open the chain can be positioned across the opening to discourage the horse from trying to jump out.

Teaching your horse to stale into a bucket held beneath him is a useful way of saving bedding.

regular has this routine become that I merely lead them in with the bucket in my hand and no sooner are they inside than they are up on their hind toes and obliging. This is not a gimmick, though I suspect that some of my friends think it so. Added up over the year it saves countless gallons of urine from being deposited onto the deep litter beds; it also gives me a chance to check that each animal is staling regularly and that the colour of what is being passed does not indicate anything unusual.

4 Keeping a horse or pony at grass

Being able to turn your horse or pony out to grass for long or short periods is helpful and beneficial to both you and your animal providing that certain rules and conditions are adhered to. But an animal at grass will not feed and look after itself; care and management are still as necessary as with the stabled horse, sometimes more so. You must see to it that a supply of clean water is available; except during the spring and summer extra feed will be needed to make up deficiencies; adequate shelter will have to be provided; and regular inspection will be required. Just any old rubbish patch will not do, though nor is very lush pasture necessary for the ordinary horse or pony. If you do not own a field of your own and can only obtain the use of a 'take it or leave it' paddock, you will have to make what improvements are needed to make it acceptable. 'Prevention is better than cure' is a saying that might well have been first thought of by a horsekeeper. Forethought and an eye that spots a possible cause of trouble are two of the best skills you can acquire when dealing with horses. When you see something that could lead to trouble never assume that it won't happen; with horses sooner or later it will. Money spent in making a paddock safe is better spent than an equivalent amount paid in vet's bills.

Choosing a paddock
Whereas wild horses and ponies manage to survive throughout the year living off their natural surroundings of moorland or prairie this does not mean that a horse or pony can similarly be turned out into a field and thereby be able to fend for itself. Wild ponies will have an almost unlimited area to roam over in order to find and select food and there will be sheltered places which together with their thick

natural coat, oily and ungroomed, will protect them through the winter.

When dealing with the food aspect of a riding horse or pony kept at grass you will need to consider two things: the size of the paddock available; and the type of grass it provides. If the animal is to be kept continuously at grass about 1¼ acres (0.5 hectares) will be required for a small pony, and larger animals will need up to double this area. This should be regarded as a minimum, but a great deal will depend on the quality of the grass and how much supplementary feeding is to be given. Where high-quality grazing is provided it will become too plentiful and lush in the spring for the average pony and it will be necessary to restrict the amount of grazing permitted in order to avoid troubles caused by overfeeding. Common problems relating to this are laminitis or founder of the hoof (an inflammation of the internal structure of the hoof) and colic caused by rich grasses fermenting in the bowels. (Incidentally, never feed your pony lawn mowings, which will almost certainly cause this latter condition.)

Ponies vary so much in breed and type that it is impossible to be more specific about feeding at grass without seeing the animal. Generally, though, the hardy types that are more able to withstand

Any old weed patch, such as this one, does not make a suitable paddock. It is also unwise to have horses or ponies turned out in areas like this where apple trees present a greedy animal with the opportunity of eating too much fruit and probably contracting colic.

bad winter weather will become too fat if allowed the same amount of spring grass as those horses and ponies that are crosses of a more highly bred character. The country or locality where you live will also have a bearing on the type and quality of the grasses available for grazing. In the United Kingdom a good quality pasture will contain a plentiful mixture of the perennial rye grasses and meadow fescues, whereas in the United States and other temperate parts of North America and in Australia there will be a wide range of various herbages from the famous blue grass of Kentucky to fescues, alfalfa and many varieties of prairie grasses. As one would need to be a botanist to identify them all, the best rule of thumb method to judge the suitability of a paddock is to look for a good mixture of sword without too much coarse and tussocky growth. Couch grass and cotton grasses are valueless as feed, as are weeds and the various types of rushes found on swampy ground. A lookout should also be kept for plants of a poisonous nature that will need to be removed before allowing the horse or pony to graze the land. Common ones that are to be found in pastures and hedgerows include ragwort, deadly nightshade and foxgloves. Hedges sometimes also contain trees and bushes that are poisonous if eaten, of which yew is particularly dangerous. I also would not allow rhododendron bushes to be reached and there will probably be others in your locality that are poisonous. Most horses and ponies find the smell and taste of these plants not to their liking but occasionally this is not the case. Poisonous plants that are eaten will cause digestive upsets, and even death results if prompt medical treatment is not given. Any animal seen to be ill that has had access to anything of a known poisonous nature should be attended by a vet as soon as possible.

I have also known ponies to have peculiar blisters or small ulcers break out over the muzzle and nose because of an allergy to certain plants present in the grass being grazed. This can happen because of the astonishing sensitivity the lips of the horse possess in foraging about to select only those plants that it finds palatable. This is especially true of the top lip, which has a large number of highly developed nerves that are used to determine the texture of the plants being grazed. Combined with the nostrils, which pick up the various scents, the horse's highly developed sense of smell and touch enable it to be highly selective in its choice of grasses. The lips are used to encompass each blade or shoot and guide it into the front incisor teeth, which then bite onto the grass and crop it off by a series of jerks and twists of the head in a snatching action that rips the grass away. The tongue then transports the food back to the molar teeth, which do the chewing. If pieces of root or dirt are torn up with the grass the lips are so sensitive that they will manoeuvre these unwanted parts to the side of the mouth, where they will be dropped out while the process of selecting and cropping fresh plants goes on. At the same time, the tongue mixes saliva with the food and the molar teeth steadily masticate it. For those not familiar with this procedure it is well worth a close inspection. Should undesirable or distasteful plants be taken back into the mouth the sensitive taste buds on the tongue identify them as such and the horse or pony will be seen to nod its head rapidly up and down to enable the whole mouthful to be disgorged and shaken out. Because of this method of grazing, which differs from that of the cow, for example, horses and ponies will generally only graze shorter grasses (up to about six inches long), often grazing very short grasses down to nothing. For this reason any paddock consisting of much long, lank grass will be less useful than one with shorter-growing grasses. A possible exception to this are meadows possessing good grasses that have been allowed to go up to seed, known as 'hay on the stalk'. Ponies will often eat this type of pasture and it will have good nutritional value. All grasses, however good, will lose their food value in the winter, a process that begins in late summer. Supplementary feeding of hay, and if the pony is working of concentrates as well, will be necessary in increasing amounts with the approach of winter.

Another point to consider when choosing a paddock is that of location. A level, well-drained pasture in the shelter of a wood or with a thick high hedge will be the best, very steep or low-lying marshy land in an exposed position is the worst. Providing proper drainage to an area subject to waterlogging will be an expensive operation, but by studying the water course a considerable improvement can often be made by digging a series of small ditches to channel the water away. Poaching of the land in the winter poses a real problem when a paddock is constantly in use. I have seen it suggested that the paddock should be divided and alternate parts used as soon as poaching makes one part bad. This may help in some cases, but on the other hand you can often extend the problem over a larger area by doing this and more good grasses will thus be destroyed. Once these are lost through trampling in the mud only weeds will grow back when conditions improve unless you renew the grass by seeding. A better method is to put down some rubble or broken stone where the horse or pony usually stands to provide harder standing and better drainage. Areas worst affected are often gateways and drinking places and where the animal is fed hay if this has to be done in the open. If you can find some old lengths of timber (the trunks of small trees are ideal) these can be placed in a rectangle on the ground where you carry out supplementary feeding, and the enclosed area can be filled with a load of gravel or even sand. This will raise its height and help to prevent poaching; it will also reduce the amount of hay lost through being trampled in the mud. If you decide to try this, do not be persuaded to put down the gravel and sand without first containing it or it will become spread and scattered and soon lost, wasting your time and money. Whether improvements of this kind are worth carrying out will also depend on whether the land is your own for your exclusive use or rented and possibly shared with others. If the latter, then make sure that you have a clear agreement with the landowner beforehand. Do not find yourself in the position of having made costly or time-consuming improvements only to find that the owner then needs the land for himself!

Before using any piece of land it is wise to go over it and remove any rubbish and dangerous pieces of material–such as old bed frames, broken gates or machinery, etc.–and also anything that an animal can get its leg through or caught up in. Lastly (and especially if it is not near to home) consider having your horse or pony branded as a deterrent against theft. In some countries this is standard practice but in others (and the UK is one of these) branding is not generally in use, though a considerable number of horse and pony thefts are regularly reported.

The other vitally important consideration is the provision of adequate drinking water. Clean, untainted water must always be available to the horse or pony. I do know of pastures having a convenient stream of clean water flowing over a stone or gravel bed with gently sloping access for the animals to enter and drink. More often, though, rivers and ponds are not suitable and can be a source of danger to the horse or pony. Steep, muddy banks and stagnant ponds should be securely fenced off, and only when pure water is present in country areas where pollution is impossible and the water can be reached in perfect safety should its use be considered. Normally the best arrangement is to provide a drinking trough with running water in a position easily reached by you as well as the horse so that cleaning it out and clearing ice from it in winter can be done easily. If you cannot provide a trough, make sure that you have access to fresh water nearby as you will have to fill buckets at least twice a day to keep the animal's thirst satisfied. Carrying the six to eight gallons (27 to 36 litres) of fresh water that your horse or pony requires each day will be hard enough work without your having to walk across an adjoining field to fetch it.

Fencing

The best kind of fencing for horses is posts and rails. It looks good and is the safest, but it is also expensive. If you can afford this type I would advise you not to paint it white, as attractive though it looks when new, it will look dilapidated and run down in a year or two unless you spend a small fortune on its maintenance. If you particularly want a small show paddock, perhaps to set off your house, there is a type of white plastic fencing that might be considered, although I am not a devotee of plastic for this use. Many firms supplying posts and rails offer them specially treated for longer life, and although this adds to the initial cost it is worth it in the long run. If you buy untreated wood, give a good coat of tar to the pointed end of the stakes and use creosote for the rest of the fencing.

When erecting such fencing some establishments prefer to keep the posts on the outside of the fence and to nail the rails through from the inside. This gives an unbroken line on the side that the horse walks or gallops along and prevents his striking the uprights. Research among vets in America, however, discovered that the incidence of injury caused to horses striking posts placed on the inside of fences was very low, whereas there were many examples of cuts and tears caused by nails projecting from rails fixed to the inside. These can stick out from worn or rotting rails nailed to the posts from the inside of the paddock. I think that there is some advantage in strength if the posts are placed on the outside of the rails, especially when an animal leans over the fence. An occasional stroll around the perimeter is in any case needed to check the fencing, and if you carry a hammer with you any offending nails can be dealt with. It is possible to purchase posts and rails that do not need to be nailed. The posts have holes cut through them and each rail has tapered ends that slot into the holes and project out the other side until they make a tight fit. The trouble with this type is that a broken rail is difficult to replace once the fence is erected, and one or perhaps two posts have to be dug up in order to fit the new rail.

Wooden rails nailed to the inside of the posts are less likely to be sprung loose by a horse's weight, but problems can arise with protruding nail-heads so a regular inspection of the fencing is essential.

A much cheaper form of posts and rails is provided by half-round posts with wooden rails made from the first cut off the tree when the timber yard begins sawing up a log to make planks. These have a flat sawn surface on one side that can be placed against the post when nailing. If collected from the yard these planks are usually available at little more than firewood prices as they are uneven in thickness and have the bark on the outer side. If reasonable widths are selected in lengths of approximately 12 ft (3.6 m) they make good fencing. Make sure that you only use pieces with sufficient thickness of wood, as ones that have been sliced too close to the bark will not stand up to the leaning weight of a horse. You will also have to remove any rough, knotty projections on which an animal could catch or knock itself. When nailing the rails to the posts be sure to drive the nail heads fully home beneath the surface of the bark: unless you do they will project when the bark falls off and you will have the problem already described.

Many other types of fencing are used by horse owners, some of which I consider unsuitable, although many people seem to get away with using them. One type that I would never use is barbed wire. In my opinion a horse or pony will eventually do itself some kind of harm on this, and if it breaks and forms a jagged rusty loop on the ground it can be very dangerous. Plain wire stretched taut and kept at sufficient height above ground to ensure that the animal cannot get its leg over it should cause no such problems. Where sheep or pig netting is used in a rented field you must see to it that this is not slack enough to lie on the ground, where a hoof could become entangled in it. If the owner of a rented field cannot be persuaded to renew dangerous parts of fencing, at least take it upon yourself (with his permission) to do what you can to improve matters.

The height that the fence should be will depend to some extent on where the paddock is situated. If it adjoins a road, or anywhere else that makes it necessary to ensure that the horse or pony cannot jump out, it will need to be at least 4 ft 6 in (1.37 m) high, and perhaps as much as a foot (0.3 m) higher. If it is surrounded by other fields where no great harm can come if the animal does jump over, then less than this will do, though under 3 ft 9 in (1.15 m) is not advisable. Best of all, of course, is the paddock that is surrounded by a good thick hedge. If this is also reasonably high it will provide protection from wind and rain as well as giving shade–protection that is vital for the horse.

A field shelter

If you intend to turn your horse or pony out during the heat of summer or the wet and windy conditions of winter you will need to provide a field shelter for its use. Choose a position that is well drained (not the bottom of a slope that will become waterlogged and then freeze over in winter) and build the shelter to as high a standard as possible. Leave one side open for easy access, and face this opening away from the direction of the prevailing wind. Make it as leakproof and free of draughty gaps as you can, and make sure that the roof is high enough. Ideally the inside ground area should be covered with rubble (but make sure that this does not contain glass or sharp objects) and topped with deep litter straw. If you can also arrange for the rubble to continue over the approach outside to prevent a quagmire in bad weather then so much the better. A hay rack and manger should also be fitted and provision made for water unless this is available close by. If you do not have any stabling and want to use the field shelter for this purpose, other fittings such as those described for stables may also be desirable. Muck out the droppings night and morning when you feed and water the animal and see to its well being. If possible, arrange the shelter so that there is a small piece of ground behind it that can be fenced off for use as a muck heap. In addition to this, a small horseproof lean-to can be built to hold a fork and wheelbarrow, bins and buckets and perhaps a week's hay ration. All this will make your life and that of your horse or pony much happier. The protection that a good field shelter gives, from flies in summer and from cold, wet or windy days in winter, will make an appreciable difference to the animal's health and condition. If a landlord will not allow a field shelter to be built or if you are unable to afford one (though I would question whether you should really own a horse at all in the latter case) it may be possible to manage without one. This will depend on the natural protection that your paddock provides and the severity of your local conditions and winter climate. In any event careful attention should be paid to the provision of additional clothing for the horse, as described below, and also to the amount of extra feed required to make up for the loss.

Additional protection

Unless your particular animal is one of the very hardy (usually pony) breeds that has a thick coat, mane and tail that can be left intact without clipping or pulling, some extra protection from the worst weather will be desirable for its comfort and well being. There is no doubt that wet and windy conditions do cause an animal a lot of misery if it is left turned out and unprotected. This is especially true when it comes to the thinner-skinned types and those that have been clipped. Don't over-groom the horse at grass, especially in winter; you should merely brush off the mud and make him look reasonably clean for riding. This will leave the natural grease in the coat that will be needed for wet days. A good New Zealand rug will also be needed – two are even better so as to keep them properly managed and dry. A heavy, rain-soaked, muddy and uncomfortable rug is useless; it will add to the horse's misery rather than alleviating it. The traditional New Zealand rug is rather heavy, being made of a waterproof macintosh type of material on the outside with a warm, wool-like inner surface. In an effort to improve on the design and make something lighter, manufacturers have produced rugs using nylon and other man-made fibres, some of which are very popular with horse owners. I think it is true to say, however, that despite its shortcomings the traditional type still finds favour, mainly because it withstands a lot of hard use, while the newer, lighter materials do tend to get caught up and tear easily. Pay particular attention to the fit of the rug; if it is too big it will slip and the animal may get entangled in it, and if it is too small extra pressure will be put on the horse's withers and the rug's front straps, causing rubbing and sore places. Keep the straps well greased to prevent them from becoming stiff and the buckles oiled to stop them rusting. Don't use a New Zealand rug at the first hint of autumn – it should not normally be needed until winter sets in. Delaying its use will encourage the horse's winter coat to grow, giving him the natural protection he needs, and the New Zealand rug should be used to supplement this. The exception to this is where owners want their animal to take part regularly in strenuous activities (hunting, for example) and have their horse or pony clipped. In this case a more judicious use of the New Zealand rug will have to be made; they should be aware of sudden changes in the weather that make the rug necessary (and the need to remove it when the temperature rises again).

Working a grass-kept horse or pony

Horses and ponies at grass that are to be worked hard must always be gradually hardened before being asked to take part in competitive events or strenuous work such as hunting. Their work and food requirement (see also chapter 5) will have to be gradually increased over a period of several weeks beforehand, depending on how long the animal has been left unworked. It is no use leaving a horse or pony turned out for weeks and then suddenly deciding to use it for galloping about or jumping at a week-end or school holidays. Used in this way an unfit animal will be subjected to discomfort and strain that at least is unkind and at most will lead to serious permanent damage. If the horse has not been ridden for some time you should begin by riding him at the walk for half an hour a day to harden his back and get him used to carrying a rider again. Increase this period of exercise by a little each day and include some level trotting to harden his legs and improve his wind. Grain, fed in the form of crushed oats, barley or concentrates such as horse and pony cubes, will have to be split into at least two feeds per day beginning with a pound or two (up to a kilo) and increasing to several pounds according to the size and type of the animal. You will soon see and feel the animal's improved condition, and once it is fit you can turn it out and leave it for a day or two between working it, or for several days after something strenuous, adjusting the amount of extra rations accordingly. The requirements of every horse vary slightly with regard to exercise, food and fitness working off grass, but if you follow a pattern of gradual change and avoid violent upheavals to the animal's routine you should not have any great problems. In winter, when returning from a day's hunting or gymkhana, etc. with an animal kept at grass, dry him off before putting on his New Zealand rug and turning him out. Feed a bran mash (see chapter 5) as soon as you can after getting home from such a day's activity and give the evening feed a couple of hours later. If for any reason this later feed cannot be given, cook an equal amount of oats or barley in with the bran and feed them together. The next day will be soon enough to brush off mud and sweat stains, when the animal should also be checked for any signs of lameness or other troubles.

Never feed a large quantity of cubes or nuts to an animal that is to be turned out immediately afterwards onto wet grass. There have been reports of animals dying from severe colic caused by the packing down of the cubes, which block the passage of the grass and cause it to ferment. Hay fed beforehand is said to prevent this because sufficient roughage is present in the stomach.

Companionship and preventing squabbles

Horses and ponies are herd animals, and it is in their nature to enjoy companionship; the turned-out animal is seldom happy on its own, especially if there are no other animals in sight. Another horse or pony that he or she gets on well with is fine; even animals such as sheep or goats will give company. If cattle are to share the horse's field they should not have horns as there is a danger of injury to the horse. Where more than one horse or pony share a paddock always be sure to feed them well apart; if feeding takes place in a field shelter it must be large enough for them not to become jealous and kick each other. The opening from the shelter onto the paddock should also be wide enough for them both to get through easily without having to crowd together or fight to establish who goes first. When catching up horses or ponies never be tempted to tie one to a handy object while you fetch the other unless it is an absolute fixture without the slightest chance of being moved or pulled over. Gates are notorious for this: a horse tied to one is in the greatest danger of dragging it open or lifting it off the hinges and then panicking and causing itself horrible injuries. Nor is it a good idea to leave a headcollar on an animal at grass that you cannot keep a constant eye on. These too can become entangled in hedges, etc., and cause the animal to cut or scratch itself.

Grass management and worm control

Another problem that arises with horses or ponies turned out onto small areas of pasture is that it will very quickly become 'horse-sick'. Infestation and build-up of worms will then result. Large establishments avoid this by resting fields that are used by horses, and also by replacing the horses with other animals such as sheep or cows, which will graze over horse droppings and not become hosts to horse worms; this helps the pasture to stay sweet. They will also possess the machinery and labour to harrow and roll their land. No such advantages will be enjoyed by the small horsekeeper, so how does he or she cope with this? Two suggestions that are often put forward I do not think are practical. One is for the horse owner to go daily over the field and pick up all the droppings. In theory this will prevent the grass from becoming sour and prevent worm larvae from hatching out and multiplying. In practice no one will find the time to do this and in any case when the weather is bad it would require superhuman strength of will to carry it out diligently. The other alternative put forward is to enlist the help of a friendly farmer to harrow and roll the field for you and perhaps top it to keep down the weeds. It is then suggested that you offer to do some work for him in return. Once again this does not relate to present-day circumstances. Farming is a skilled and highly mechanised business and farmers today cannot afford to have expensive labour and machinery employed doing odd jobs as favours. Likewise, any help that you could give him, even supposing you had the time, would probably be more of a hindrance than a help.

A handy, easy to use rough grass cutter for small patches of weeds or for long grass. A metal disc fitted with removable cutting blades is mounted between the wheels and driven by an engine directly above it; the machine is merely pushed through the undergrowth. An inexpensive form of post and rail fencing can be seen in the background.

My own method is to use a thirty-year-old tractor to which I attach an old, reconditioned rotary mower that works very successfully, topping tall grass and weeds before they can seed. I also use an old set of harrows to spread and scatter the dung piles and dead grass, which gives the fresh grass a chance to come through. For those without heavy machinery of any kind there are still several things that can be done to assist management. Walk around the paddock wearing rubber boots, perhaps with some friends or children to help, and kick over and scatter the dung piles. This will allow the air and sunlight to get to the grass patches beneath and also enables birds to get to work on the worm larvae. If you carry a small scythe or stout stick you will be able to knock over tall weeds at the same time; this is also a good moment to check your fences. Unless the paddock is very small it is also worth considering whether it could be divided so that half can be rested after this treatment. For those able to afford a small grass-cutting machine there are several types available that are designed to cope with rough grass and weeds. The one that I would recommend is the type that has a rotary cutter driven by a small petrol engine mounted between two large pram-type wheels. This is very manoeuvrable, can be pushed close to hedges and trees and can also be used on banks for clearing weeds. On a fairly low setting it will also scatter piles of droppings.

Grazing horses will also deplete the lime content of the land, and it is prudent to replace this at least to some extent every few years. I have seen this lime requirement given as several tons per acre, but few people will have the means at their disposal to handle this quantity. Half a loaf being better than none, it should be possible to slit open a few bags of lime and scatter it downwind with a shovel. If you do this in the autumn the winter rains will disperse it for you over a larger area. One man that I know reckons that as his neighbour regularly limes his fields and they are situated on sloping land above those of my friend, he doesn't need to lime his own land at all, as the rain gives the benefit to his lower fields by washing the lime down. So if your field is on a slope spread what lime you use over the upper part of it.

To sum up, horses and ponies kept at grass in the way described here are often as fit and healthy as their stabled counterparts and sometimes more so. This is particularly true of the tougher breeds of pony, though the larger, thinner skinned horses tend to do less well, especially if all the needs outlined are not met.

Winter snow and severe conditions

Some countries will be subject to very severe winter weather for long periods that make it impossible to keep a horse or pony outside. It will also make exercising the animal a problem where no indoor facilities are available. Owners without such facilities but with a barn, such as those in America that house loose-boxes with a central alley, can use this to walk or ride the animal to provide some exercise and prevent boredom. If deep snow covers a known area it can be partly cleared and then ridden through to provide a rectangle large enough to do some exercise, and in this way you can generally get through the worst months and be able to keep your horse or pony well and happy. When the snow is only up to about 2 ft (60 cm) or so deep it can be ridden through; it can be exhilarating as long as you keep to places you know that do not have hidden hazards covered by the snow. In Switzerland, where these conditions exist every winter, it does not prevent horses being exercised outside most of the time and the horses really enjoy trotting or cantering through deep snow when it is loose and powdery. Wet snow will pack itself into the hoof, forming a 'ball' on the bottom of the foot that can become surprising difficult to dislodge as it freezes and forms a solid mass. The horse will then become unbalanced and stumble until it breaks away. This can usually be prevented from forming in the hoof by an application of grease or melted-down fat that should be worked well into the hooves covering the sole of the foot and the frog. Frozen snow and ice cause bigger problems when it comes to exercising but even here it is possible to do some work outside by having spikes or studded shoes fitted to the horse. When the weather is dry it will do no harm to let your horse or pony spend some time during the day out in a paddock that is snow-covered – in fact you will find that most of them enjoy playing and rolling in it. Although animals will dig through the snow and find the grass and roots beneath to eat, any horse or pony turned out in these conditions will still have to be fed a full ration of hay and concentrates to supply its feed requirements.

An ideal riding horse, with elegance and strength: a nice head on a strong body and legs, with well-defined withers, good sloping shoulders and high, powerful quarters.

The author's stables, which as can be seen are sited close to the house. The stabling is constructed from tarred wood on a concrete base, with a pointed roof clad with clay tiles and fitted with ventilation louvres; this is an ideal arrangement.

A good example of the type of manufactured loose-box that can be brought to the site and erected on a concrete base.

Feed room, showing the galvanised metal bins that are vermin-proof and long-lasting. The blackboard is used to record dates of shoeing and injections, and to provide details of the feeding requirements of each horse.

Back doors to the loose-boxes allows mucking out to be done under the covered area at the rear. Note that the level on which the wheelbarrow is standing is below the floor level of the box, making the work easier and also giving more head-room under the covered way; deep-littered boxes can be cleaned out straight into the trailer. Having your tools and equipment stored under this covered way keeps things handy as well as tidy, and enough hay and straw can be stored under cover here to last for a week or more.

Ragwort, a common poisonous plant that is often seen along hedgerows. Any that is present in a paddock intended for horses must be dug out and burned. A lookout should also be kept for ragwort that may have been baled with meadow hay.

A separate small exercise paddock, useful both with stabled horses and for ponies kept at grass.

The right way to provide food and drink for a grass-kept horse: the hay-net is well up out of harm's way, the hanging manger and water bucket are also at the correct height. Be sure to provide one set of containers for each horse, and to keep them well apart to prevent jealousy.

A hay-net being weighed on a spring-balance. The hay-net filler on the wall hinges up into a horizontal position, and small hooks keep the neck of the hay-net open while it is being filled.

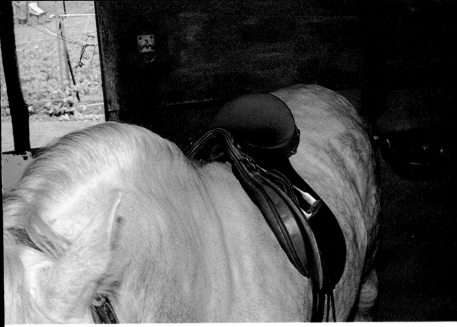

A medium-size, spring-tree general purpose saddle. When fitting a saddle it is very important to ensure that no direct pressure is put upon the horse's spine: (Above) shows the cut-back head, which gives good clearance over the withers; in (Below) the central channel over the spine is clearly shown.

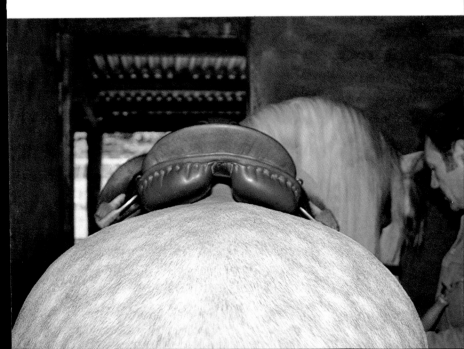

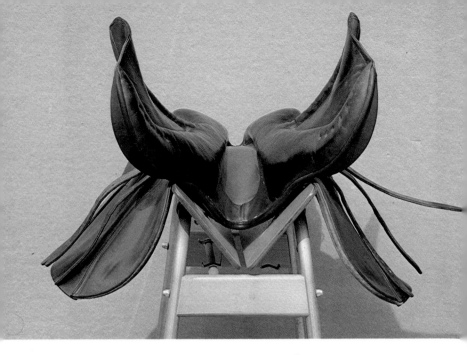

A useful saddle stand. In (Above) it has been adjusted to form a vee that holds the saddle so that its inside surfaces can be cleaned easily. In its normal position (Below) it can be used to clean the upper surfaces and also to store the saddle.

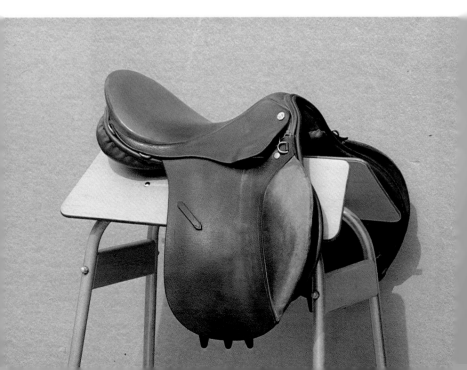

5 Feeding requirements

The correct feeding of a horse or pony is both easy and difficult. How can that be? It is easy because if you stick to the basic principles of turning a horse out to grass, or feeding a stabled horse mainly hay with some additional oats or cubes when at work, the animal will generally do well enough. It is difficult because every horse is an individual when it comes to feeding and keeping it in perfect condition is an art rather than a science. It would take the whole of this book to go into the theories because the horse, unlike many other animals, has had comparatively little scientific research done about its feeding habits or requirements. From work that has been done on this some principles could be listed–protein and starch equivalents of various foods and other technical details–but for the average owner wishing only to keep an ordinary horse or pony for normal activities, technical information of this kind is more of a hindrance than a help. (I have seen charts published that have confused me, let alone a person wanting to know what to feed for the first time.) I will therefore confine my advice to some easily understandable do's and don'ts, together with some basic facts about feeding. If these are followed you will not go far wrong to begin with and will soon learn what type of food and feeding best suits your particular animal according to the work it is doing.

The stomach of the horse is small when compared to that of, say, a cow–about 2 gallons (9 litres) capacity as against 44 gallons (200 litres) in the cow. Food quickly passes through the horse's stomach to the rest of the digestive tract. It is in the caecum, or second stomach, which is in fact an intestinal canal that holds about six or seven gallons, that digestion takes place. For this reason the horse chews rather slowly, and in order to feel satisfied needs to have a sufficient amount of bulk in the form of hay or grass. There is a 'complete' food in the form of manufactured pellets, which provide the horse with all its actual nutritional requirements, but in my opinion this is not a good method for the simple reason that it does not supply enough bulk to keep the animal content.

As a guide to correct feeding I have included a chart (see Fig. 4) that takes into account the need for sufficient bulk when feeding the stabled horse. This chart should only be used to start you off in finding what best suits your particular horse or pony, and should not be taken as strictly accurate in every case. When using it for a stabled horse that is spending some time out at grass, allowances should be made to take into account the feed value of the grazing. For example, if in summer a horse normally stabled is turned out to graze for two or three hours after its morning exercise, then the

SIZE h.h.	INTAKE bulk required daily lb (kg)	HAY lb (kg)		CONCENTRATES cubes, oats or barley lb (kg)		BRAN lb (kg)
13.2	15 (6.7)	a.m.	3 (1.35)			
		noon	3 (1.35)	1½	(0.67)	½ (0.22)
		p.m.	5 (2.26)	1½	(0.67)	½ (0.22)
14.2	20 (9)	a.m.	4 (1.8)			
		noon	4 (1.8)	2	(0.9)	½ (0.22)
		p.m.	7 (3.17)	2	(0.9)	½ (0.22)
15.2 (light horse)	25 (11)	a.m.	4 (1.8)			
		noon	5 (2.26)	3	(1.3)	½ (0.22)
		p.m.	9 (4)	3	(1.3)	½ (0.22)
15.2 (heavier type)	29 (13)	a.m.	4 (1.8)			
		noon	6 (2.7)	3	(1.3)	½ (0.22)
		p.m.	12 (5.4)	3	(1.3)	½ (0.22)
16 plus	34 (16.5)	a.m.	5 (2.26)			
		noon	6 (2.7)	4	(1.8)	½ (0.22)
		p.m.	14 (6.34)	4	(1.8)	½ (0.22)

Fig. 4 The chart provides a general guide to the feeding requirements of a stabled animal in light work; if hard work is performed more feed in the form of concentrates will be required. The amount of bran fed should only be marginally increased, however, as it not only has a high protein content but is also rich in phosphorus, and too much of it will upset the mineral balance.

midday feed of hay and concentrates should be omitted. If experience proved that this was taking too much away from the animal, some of it could then be added to the later feed given in the stable. Used in this way the chart should provide a basis for correct feeding, probably erring on the side of slight under-feeding, which will have less adverse effects than over-feeding. Some people may like to vary the method of feeding the suggested concentrates by including some of this with the morning hay feed, or to split the evening hay ration so as to provide an extra late feed. The amounts suggested in the chart cover maintenance and some light work such as a short hack or twenty minutes' lungeing, so if the animal is being worked hard extra food in the form of concentrates will be needed. Similarly, if your horse becomes sluggish or loses condition on the amounts given, then this will also call for increased rations until you find the right balance to keep your horse fit without becoming over-excited. To begin with, concentrates can be provided by using horse and pony nuts or cubes, but later these can be replaced with crushed barley and oats if you wish. This will be cheaper and you will also know exactly what ingredients are being fed – something I never feel sure about with manufactured foodstuffs.

Hay and grains

Hay less than six months old is too young to feed to horses; after two years – earlier than this if it is not properly kept – it will be losing its feed value. Bad hay that is musty or mouldy should never be fed, as apart from being low in feed value it is very harmful to the lungs and stomach. Recognising bad hay is easy enough, but judging the feed value between average hay and that which is very good is more difficult. Some say that good hay should smell sweet, others that there should be no smell at all; sometimes one is advised to look for slightly green-coloured hay, sometimes bright yellow or even blue-coloured. Hay that has lost its colour is said to be no good. Then there are the different kinds of hay: 'hard' or seed hay, 'soft' or meadow hay; and in some countries, lucerne hay or alfalfa. What does all this mean, and what does the average horse or pony owner do to get it right?

The colour and smell of hay will depend on the grasses it contains, the way it has been made and stored and its age. It should never smell bad – musty or sooty – or have mould or spores on it. Break open a bale at random and try this. If it smells sweet or has no smell at all then it is all right. Sometimes hay that has 'cooked' a bit in storage, owing to a period of heating, will smell slightly of mushrooms. If there are no other bad signs don't be put off by this, as horses often like such hay and do well on it. In fact some horses will readily eat hay that seems to have a slightly 'off' smell in preference to a sweet-smelling variety. I cannot explain this, and to add to the mystery even bales of hay from the same field will sometimes be treated in this way. Before buying a large load, therefore, it is prudent to try two or three bales from different parts of the stack to make sure that your horse will eat it. It may be, in these days of crop spraying and spray drift, that some parts of a field have been affected, and although not distinguishable by humans the horse's keen sense of smell detects it.

As to colour, if hay is black discard it; if it is very pale it is probably too old and will have lost a great deal of its food value. Green, sappy hay, on the other hand, is too young and can cause digestive troubles if it is fed before it has become properly cured. Hard hay or seed hay is specially sown to be cropped as hay and is often composed of rye grass and other good quality grasses high in protein. For this reason it is usually much sought after by racing stables. Lucerne or alfalfa makes excellent hay, also with a very high protein content. In countries where this herbage can be grown it can also be cut and fed green after it has been allowed to wilt in the sun for a few hours. As many as six to eight cuts a year are not unusual in areas where the climate favours its growth. Less high in protein value is the soft hay made from grass meadows that are grazed for part of the year and then 'shut up' to be used for hay. This is considered to be better for ponies, the more common-bred horses or for sick animals, as it is less rich and is more easily digested than the coarser types. It is usually more easily obtained than seed hay and is therefore cheaper. If it is well made it can, however, cost nearly as much as the hard hays; in years when hay is in good supply don't buy a load that is cheap, only to find when you come to use it that it is full of weeds.

Most authoritative books on horse feeding will insist that the reader only feeds the finest hay and grain such as oats. This is sound advice but unfortunately not always an easy thing to do. One is sometimes given the impression that the farmer or merchant will offer a selection from the best, most perfectly made hay and the plumpest oats that 'rattle' when one's hand is pushed into the sack. This harps back to the old days of a buyer's market. Obviously you must try to get the very best you can, but occasionally you will have to take what you can get or go without. In difficult years, good hay that has been well made can be hard to come by and in some areas almost impossible. Even if you are desperate never use hay that is black or mildewy, and if you really have to use dusty hay damp it before feeding it. If, in a very bad year, you run out of hay and cannot obtain a reasonable quality, cut down your horse's work and get the best quality barley or oat straw and substitute this, preferably as chaff. If mixed with an easily digestible food such as molassine meal (manufacturers have various names for this product, which is made from the residue of refined sugar beet), this chaff will be more tempting for the horse or pony to eat. This will get you through the worst period until you can locate some hay that you are able to feed.

During periods of enforced rest, and especially when hay is in short supply, the animal will need to have its diet supplemented with an energy-giving food that at the same time does not over-excite him or tax his system. This is where barley is a more useful grain than oats. Barley is less toxic than oats and can be fed to an unworked stabled horse in small quantities up to several pounds (say, 2 kg) per day without it causing harmful side-effects. Fed boiled it will help to keep flesh on the horse until exercise can be resumed and feeding is returned to normal. If oats are fed to a horse or pony that is not in work their high toxin content will be more than the animal's system can cope with. This will result in various troubles, the most common of which is 'filled legs'. For a fit horse in hard or fast work, it has been proved that the vast amounts of energy it needs can be best provided by feeding oats as the concentrate. Unfortunately many horses, and more especially ponies, become too 'hot' and excitable when given fairly large quantities of this grain. For the riding horse or pony, therefore, I believe that it is better to feed good hay and crushed barley, together with appetising quantities of molassine meal and carrots as the basic ration, and only feed oats for the increase in food intake that is necessary when working hard.

All grains should be fed rolled, bruised or crushed, as this makes it much more easily digestible and also prevents any of it from passing through the animal still in its whole form, when no food benefit will be obtained. Once the outer shell of the grain has been broken, however, its food value begins to diminish, so never buy more than can be used in two weeks or so. Fed alone, oats are more easily digested than barley as they have a higher proportion of husk and the outer layer is less hard. This extra husk helps to prevent the feed from packing down in the horse's stomach, as can happen with grains if they are fed alone. In the old days, chaff or chopped hay was added as roughage to prevent this packing down, but chaff-cutting is a time-consuming business and chaff is rarely used today. It is unnecessary if hay is fed to the horse while the concentrates feed is being prepared, which will provide some roughage before the concentrates are eaten, after which the animal will go back to the hay and munch away in contentment to the benefit of its digestive tract and psychological well being.

Other foods and feeding requirements

Horse and pony nuts or cubes are a useful way to feed concentrates when given in conjunction with hay as they are carefully balanced by manufacturers to include the right amount of fibre together with carbohydrates, proteins and minerals. I have already stated that I have reservations about knowing exactly what goes into them but it must be said that nearly all horses like them and eat them readily. Although more expensive than oats, etc., the owner with only one or two horses to feed may find this is worthwhile because of their convenience and the fact that they keep better if properly stored. They do not have an unlimited life, however, and one should try to buy them from a retailer with a good regular turnover. Old stock will be less palatable even if the feed value is still largely intact. On several occasions I have had them refused by horses even though they seemed all right; subsequent examination by the manufacturer found that the milk powder used as a binding agent had gone sour. Always treat with suspicion any that are found to be soft or sticky and ask the supplier to check that the batch number is still in date. On the other side of the coin it can be argued that they provide a way of feeding that is not affected by poor harvests and bad weather that make it difficult to obtain good samples of individual grains such as oats and barley. Various types of nuts are available – such as 'horse and pony', 'racehorse' and 'stud' qualities – the first named will probably suit readers best. Use them as I have outlined or according to the maker's instructions; don't feed them indiscriminately without hay; and only feed a little, if any, to a resting or sick horse.

For sick horses or during winter to provide warmth and keep weight on, boiled barley is a useful food. The whole grain of barley should be used for this; ½ lb to 1 lb (0.22 to 0.45 kg), to which one or two pints (between half and one litre) of water is added, will be sufficient for one feed. (If feeding more than this add chaff or feed hay first to provide roughage.) Used as a supplement to other food for the working horse it can be added to the feed when it is cooked. Cooking consists of bringing it to the boil and then letting it simmer for approximately four hours. When ready it will be a jelly-like substance (the husks don't dissolve) and it can then be mixed with crushed barley, dry bran and/or molassine meal to take away its stickiness. It can be allowed to cool completely before being given, or used as a warm feed on wet winter days, when it will be particularly welcomed.

Sugar beet pulp, sold in 112 lb (50 kg) bags of compressed dry cubes, is also a useful food, especially as a winter supplement. These cubes must be soaked and will absorb three times their own volume of water. Manufacturers stipulate that this should be done for a full twenty-four hours in order to ensure that the cubes are fully soaked before being fed. There is no doubt that terrific expansion needs to take place, and if this is not done with prior soaking it could very easily have fatal consequences to an animal. However, I have found that it is safe to use them after soaking for eighteen hours, which allows those soaked overnight to be fed the following lunchtime. Left for more than several hours beyond this they can very quickly go sour, so always use up those that have been soaked overnight by the end of the following day. As with boiled barley, they are best fed mixed with other dry food to absorb their wetness.

Bran is a useful food both for feeding dry and in a mash. It does not keep very well: it has a tendency to absorb moisture, and if it is kept too airtight it will also heat up. It therefore pays not to keep too large a quantity, but for those owning just one horse or pony and therefore not using vast amounts, this can be difficult. Some merchants will supply 56 lb (25 kg) bags but mostly nothing less than 112 lb (50 kg) is obtainable. Sometimes it is possible to find a pony owner nearby who has the same problem and will therefore share a large bag with you. Large, flaked, broad bran keeps the best and is best for horses, but here again you will have to accept whatever your merchant stocks. A bran mash given to a sick horse or one that has had a hard day is a well-proven and very good feed. It is also good practice to feed it once a week to a stabled horse having concentrates, as it is mildly laxative and clears the system. (Horses out at grass need no such laxative or cleansing action.) The ideal method is to feed the mash the night before a rest day, when of course no concentrates will be fed to the horse. To make the mash you will need a clean galvanised bucket, a stout cane and an old towel, together with the bran and one or two kettlefulls of boiling water. Half to threequarters of the bucket should be filled with bran according to the size of the animal, and the boiling water poured onto it. Never fill the bucket with more than this as it will be wasted when you stir it. Stir it with the cane to ensure that the boiling water is well absorbed and that no parts are left dry. It should not be overwetted but given enough water to make the mixture crumbly and binding. Put the towel over the top, leave it to cook for between twenty minutes and half an hour, and then let it cool down to a comfortable eating temperature before feeding it to the horse. Some horses relish it so much that if you feed it too hot they will burn their mouths trying to eat it. I always add a sprinkling of salt, and to increase the feed and cool it down you can add a handful of crushed barley or oats. Molassine meal and sliced carrots will also make it very tempting, especially to a sick horse off its food. Any of the mash that is not eaten up within the hour should be removed from the manger as it will quickly go sour. You should also clean out the mixing bucket as soon as the mash is fed, as later the small particles that remain sticking to the side of the bucket will be very difficult to remove.

Carrots are a welcome addition to the diet when in season and not too expensive. One is always advised to feed these sliced rather than chopped to avoid the risk of choking. I have never seen the logic in this advice. Horse and pony cubes are fed, apples are quartered and given without ill effect, and yet we are told that carrots must be sliced. Personally I always whittle mine with a sharp knife straight into the feed, for no other reason than it is easy to do and because it makes them go a long way – the horse enjoys their taste throughout the meal.

There are several other root crops and foodstuffs that it is possible to feed to horses and ponies, many of which are used by manufacturers for their products, but I do not intend to go through the list. Many of them are not easily obtainable by the average owner, and to recommend their use in advice on good basic feeding would, I think, only serve to confuse the issue. However, there are some that are worthy of mention for special purposes. Boiled linseed is often used to good effect for improving the condition of sick or convalescing animals. It also has an improving effect on the appearance of the coat. A tablespoon of cod liver oil in the feed (if the horse does not object to the smell and taste) will also help to give a dull coat a lustre. Molasses is also helpful as an additive in countries where it is generally available, and to help promote strong healthy growth of the hoof gelatine can be added to the feed. Generally speaking, though, if your animal is getting the correct amount of the basic foods as outlined in this chapter there should be little or no need to worry about extra additives. If in doubt, especially during winter, buy a vitamin supplement or one that addresses itself to your particular requirement and use this to see if you gain the desired improvement.

Two other important items that should be freely available to the horse or pony are good clean water and salt. A stabled horse will require between six and eight gallons of fresh water per day, and if this is not provided by an automatic drinker then regular topping-up of the water container will be necessary. Salt should always be present in the form of a salt – or better still a salt and mineral – lick. A certain amount of salt will be present in most foods, but it is good practice to provide a block or 'brick', and an easy method of doing this is to simply place one permanently in the manger.

Digestive problems

Never suddenly change the diet or feeding routine of your horse. To do so will nearly always cause him discomfort, if not actual harm. Make any changes gradually, over a period of a few days at least and preferably longer, so that his digestive system can become accustomed to the change. An exception to this is where you are suddenly forced to interrupt the regular feeding and exercise pattern for your horse because of injury or very bad weather, and then only feed hay to the stabled animal. I have known very fit racehorses who were full of concentrated food to be suddenly denied the exercise they needed to clear their system and die as a result, despite immediate steps being taken to prevent their being given any further amounts of highly energising food. It is doubtful whether readers will be faced with quite such an acute problem as this, but if in doubt a vet should be consulted. He can administer an injection to relax the muscles of the bowel and also employ other measures to help the animal through this period. Similarly, if you witness your horse or pony behaving in a manner that suggests colic (a broad term for stomach and digestive disorders) a vet will be needed if the animal does not improve within an hour or so. Signs that all is not well are when the animal is unduly restless, frequently turns his head to look at his stomach and makes attempts to kick his sides and stomach, gets down and immediately gets up again, or attempts to pass water or droppings without much success. Providing these symptoms do not become violent, lead the animal out for a short walk round and often this will relieve the condition. Withdraw all food but allow the animal to drink if it wants to. If the condition cannot be relieved in this way stay with the animal and try to keep him on his feet until the summoned vet has arrived.

6 Tack and other equipment

'You only get what you pay for,' is an often repeated saying that rings particularly true when applied to horse tack. By the same token, you pay a lot more for a little extra once you reach a certain level of quality. It is well worth looking and comparing quality and prices of tack before you buy your horse or pony, so that you have a good idea beforehand where the best value is to be found. Don't leave this until the animal arrives so that you then have to rush out and buy whatever is available from the nearest supplier, even if it costs the earth! Of course, if the horse's own tack is for sale together with the animal you are buying, this will save you a lot of fitting problems. Provided that it is a good fit and the owner does not want too much for it, it is well worth buying even if you replace some parts of it later for something better. Remember, too, that by keeping your tack clean and storing it properly when not in use it will look better and last longer; never, therefore, drop the bridle in a heap or leave the saddle in a 'splayed' position on the floor.

Saddles

The most important acquisition for your own comfort and that of the horse is the saddle. Other items are important but will be far less costly to replace if you make a mistake. Saddles come in a variety of shapes and sizes to fit you and the horse, as well as different types for various purposes. There is not the space in this book to deal with them all, so I will briefly explain some of these and concentrate on what is most likely to be used by the reader.

Apart from the American Western saddle and similar variations such as the Australian and Spanish saddles that are designed for a specific purpose, there are three types of English saddle now in use: the jumping saddle, general purpose (sometimes called all purpose) saddle and the dressage saddle. Each of these is slightly different in design. The jumping saddle is made to position the weight of the rider further forward over his knees; the dressage saddle positions the rider's weight farther back, with the legs carried longer and lower; the general purpose (GP) comes somewhere between these two. Children's saddles usually resemble this last category. It is important to remember that while a GP saddle can be used for all three purposes with some degree of success, it does not mean that the one saddle will fit a multitude of different shaped horses and ponies. Readers will best consider a GP saddle of leather with leather panels (the part that bears on the horse). Don't try to save money by acquiring a stiff inflexible saddle that will rub both horse and rider, but look for a saddle with leather that will not take too long to soften and be comfortable. It should be of 'spring tree' construction, which makes it more resilient to the seat and also ensures that the saddle head is set back at an angle to give good clearance over the horse's withers. The exception to this is children's saddles, which because of their smaller size are made without springs; they are known as 'rigid tree' saddles. The 'bars', which carry the stirrup leathers and thus the rider's weight, should be of forged steel as opposed to cast steel, which is less strong. Where forged steel has been used this is usually stamped in the metal.

English saddlers still fit these bars with a thumbcatch. This is designed to keep the stirrup leathers on the bar when in the up position, and in theory springs open to release the leather if the rider falls and is being dragged. As this catch is in my experience always left in the 'open' position, I feel that it is about time manufacturers redesigned the bars. The Australians have a better system: the bar is given a curved end thus achieving the desired result without using a mechanical device.

The saddle is built over the tree, which provides the arch that gives clearance over the horse's backbone; narrow, medium and wide fittings are available. The choice of width will depend on your horse, but a medium tree fits by far the greatest number of horses. With the rider on it, the saddle must be clear of the top of the withers and must also not rest upon or pinch the horse's backbone. Apart from this it must fit as evenly as possible over the back, thus spreading the rider's weight without causing pressure points. Allowance must be made for new saddles, which tend to 'perch up' but will quickly settle down when in use. As a rider you will find this perched up feeling quite noticeable at first, but if it is properly made the

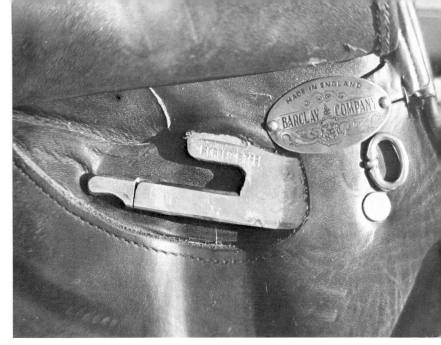

It is important that the bars of the saddle, which carry the stirrup leathers, are stamped to show that they are made of forged steel. Note the (generally redundant) thumbcatch on the end of the bar.

saddle will soon mould itself to your shape and that of the horse.

If you buy second-hand the advantage – apart from the cost – is that the saddle will be broken in and will feel more comfortable. Pay particular attention, though, to the stuffing to see that it has not become hard or flattened. Never use a numnah to try and make a badly fitting saddle fit better.

The length of saddle needed will depend on your seat and the length of the horse's back. Adult sizes range from 16 in to 18 in (40 cm to 45 cm), measured from the stud on the pommel to the top edge of the cantle. Smaller sizes, suitable for children, usually start at about 14 in (35 cm) and go up to 16 in (40 cm). If in doubt go slightly larger rather than smaller – it is no use having the saucer smaller than the cup!

Stirrups

Stirrup irons come in a variety of shapes and sizes, but I do not think one needs to look beyond a plain iron of the type illustrated. These are now mostly made of stainless steel, and fitting rubber treads to them makes it very much easier for the rider to maintain a correct

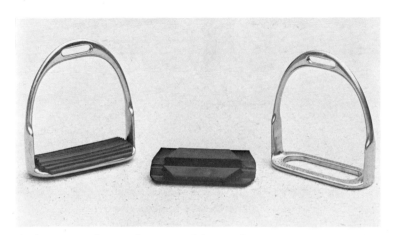

Ordinary stainless steel stirrups of the type known in England as plain hunting stirrups (irons). The rubber tread shown helps to provide a firm, non-slip surface for the rider's boot, though the stirrups themselves have a roughened platform that can also be seen (right).

contact on the ball of the foot. Buy stirrups wide enough to allow between ½ to 1 in (1.25 to 2.5 cm) space on each side of the platform once the boot is resting on it. Less than this will greatly increase the danger of a foot becoming trapped should a fall take place, and more than this increases the possibility of the foot going right through. The height of the stirrup arch is also a factor to consider: too low, and it could trap the toe; too high, and the foot could go right through. If the stirrup is both too wide and the arch too high,

This stirrup is the right size in relation to the boot shown: the height above the foot gives enough, but not too much, clearance, and the width allows sufficient space on either side of the sole without being too close-fitting.

the chances of this latter happening are obviously much greater – a point worth remembering with child riders. Good, heavyweight stirrups are best as these free the foot most readily in a fall. The stirrup leathers should be the best you can afford as they will be required to accept a lot of strain over the years. Leathers made from cowhide, rawhide or buffalo hide are the toughest, particularly the last of these. New ones will stretch, so as most people tend to ride unevenly it is a good idea to change each side over occasionally to keep this stretching even. Always mounting the horse from the left-hand side will also cause the leathers to stretch unevenly. Inspect second-hand ones very carefully for wear to the buckles and stitching, and for cuts or excessive thinness where the irons have been held.

Girths

There are many patterns and materials to choose from, but I will confine my list of possible choices to three, all for use with English-type saddles. The first, and most expensive, is leather. The Balding, Atherstone and threefold girths are good examples, but each of these needs to be given plenty of dressing with oil to get and keep it supple. When new they can rub and gall a horse very easily owing to their stiffness. This is especially true of the first two named. Money can be saved here; both less expensive and very kind to the horse is the lampwick girth. This is a fabric girth that is very soft and needs nothing more than a good scrub to remove dirt and sweat, plus attention to the leather ends attaching the buckles. I have heard it criticised for its tendency to break suddenly, but I have never personally experienced this. Cheapest of all, and very effective, are

Three types of girth: the leather Balding (top), the lampwick, made from a tubular fabric material (centre), and a nylon cord girth (bottom). Prices for these different types will be in the same order.

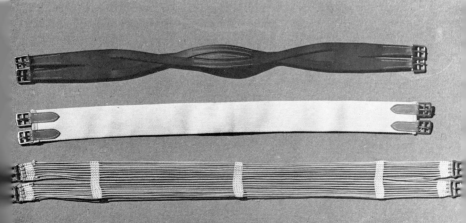

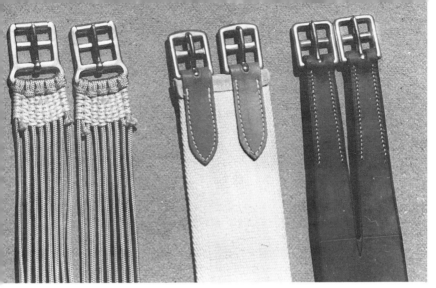

Nylon girth buckles (left) tend to be rather large in order to accommodate the cords, which make for a greater lump under the saddle; the rollers fitted to the top bar may also cut the girth strap if the metal opens. The other two examples are neater, and have grooves in the top bar into which the tongues fit.

girths made of nylon. My own complaint about these is that they are often fitted with inferior buckles.

Whichever type of girth you choose, buy one that fits just above midway up the girth straps for normal use. Then if the horse or pony is fat through coming up from grass, or if a numnah is fitted, the girth will still be big enough. In order to prevent the girth buckles from wearing a hole in the saddle flap, leather 'safes' are sometimes fitted. These slot onto the straps of the saddle that hold the girth and rest on top of the buckles. I do not like them, as they make a bulky lump under the leg, though they do save wear.

Bridles

Some of these can be very costly, especially those with fancy stitching or embossed leatherwork. Plain leather bridles are adequate and less expensive, but when finding a reasonably priced one make sure that the size has not been skimped. Three sizes are available – pony, cob and full size. Many that are sold as full size are not; likewise, some cob sizes are only large enough for a pony. Make sure that the browband is long enough to allow the headpiece to fit behind the ears of your horse or pony with enough room to prevent

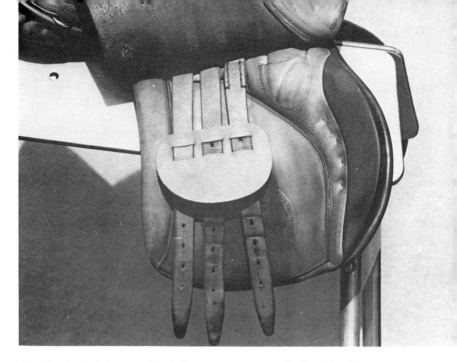

Leather 'safes' fit onto the girth straps to prevent the buckles from wearing a hole in the saddle flap. (Above) This type, which slips over the straps, causes an additional lump under the flap; the kind that is sewn onto the panel where the straps are attached (Below) provides a buckle guard that is less bulky and fits more smoothly.

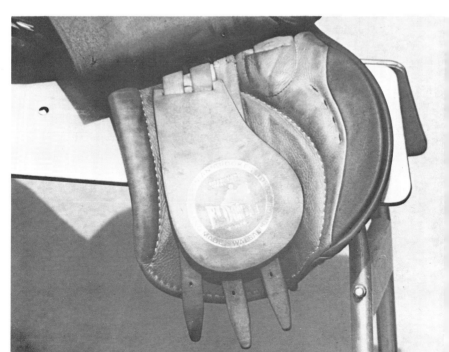

any pinching or rubbing; the noseband should have straps that are long enough for it to be adjusted just below the cheek bones and the band itself loose enough when buckled to allow two fingers to be inserted beneath the jaw; the throatlash should also be long enough to allow four fingers to be inserted when it is fastened, so that the horse is not restricted when moving its head forward; finally there should be adequate length in the cheek-pieces to allow for bit adjustment.

Reins

Plain leather reins give a good 'feel' on the horse's mouth, but become very slippery in wet weather and make a sweating horse almost impossible to hold if the animal pulls. Better in this respect are the rubber-covered reins, though these do tend to be rather bulky for normal use. I prefer plaited leather reins (or the laced type, which are similar) if I am riding a difficult horse or during rainy weather. To be avoided are the nylon plaited type; these become horribly slippery when wet, and when dry – especially if the nylon is worn – will badly lacerate the fingers.

Bits

Stainless steel is the material widely used for bits and usually lives up to its good name. Some English manufacturers have ceased making them as they cannot compete for price with those imported, many of which are now of high quality. Bits made from solid nickel do not rust and are cheap, but they are not to be recommended as they turn yellow and bend and break fairly easily. Nickel mixtures made under specific trade names cost more, keep their colour and rarely break.

A plain leather bridle suitable for everyday use and also used for hunting. Note that the noseband headstrap is slotted into the noseband itself; an alternative method is for this to be sewn onto the noseband, but it wears more quickly and also allows the noseband to droop and lose its shape. The reins are of the laced leather type.

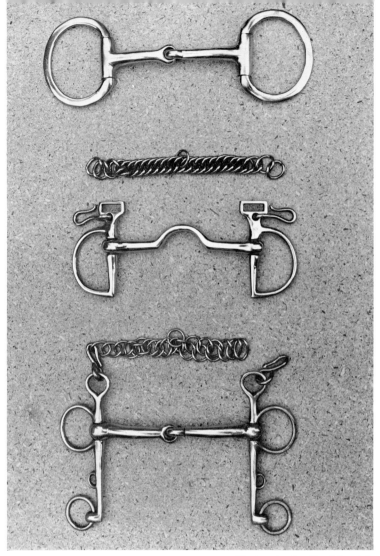

Three popular types of bit: the eggbutt snaffle (top), the Kimblewick (centre) and the jointed Pelham (bottom). To measure a bit, lay it flat and take the measurement between the cheeks – this gives the length of the piece that is suspended in the horse's mouth.

The size of the bit is measured between the rings or the cheeks, that is to say, the size given is that of the part that goes into the mouth between the parts each side that prevent it from sliding through. Sizes range from 4½ in (11.5 cm) for ponies up to about

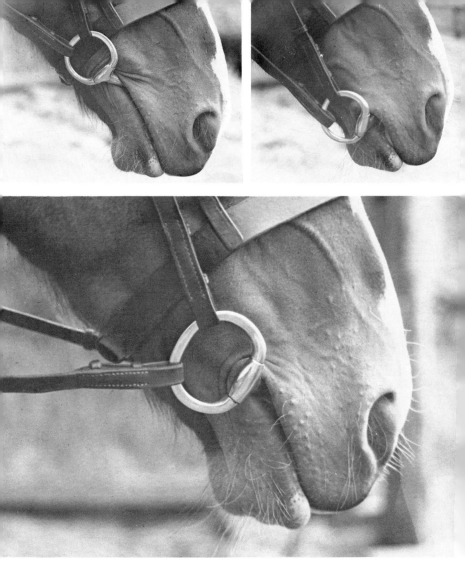

Fitting the bridle. (Top left) An incorrectly fitted bridle: this one is either too small or has been adjusted too tightly, pulling the bit up too far in the mouth and over-puckering the horse's lips. (Top right) Again incorrect: the bridle is adjusted too loosely, allowing the bit to hang low in the mouth, where the horse can get his tongue over it, and it will also strike against the front teeth. (Above) Correct: the bit is positioned in the mouth to give good contact, and the horse is comfortable and relaxed.

5¾ in (14.5 cm) for the larger horses. Don't get one that pinches the animal's mouth because it is too narrow, nor one that slops from side to side because it is too wide.

For the type of animal likely to be acquired by the reader, I have restricted the type of bit dealt with to three kinds, as will be seen in the illustrations. The one most often employed is the snaffle; the eggbutt is very popular, and variations of this with either 'dee' cheeks or loose rings are also often used. Most horses will go well in this type of bit. The Kimblewick has a stronger action, especially when used with the curb chain. It is useful to help a child control a wilful pony or for hunting a strong horse. I have seen it criticised as being too severe, even to the point of being capable of breaking a horse's jaw. While I do not believe this, care should be taken, especially with a child, that violent tugs are not given to the horse's mouth as this will cause bruising. The jointed Pelham combines a nutcracker action with a curb pressure, and for those able and wanting to use two reins will prove effective. It should be remembered that bits with long cheeks or shanks to which the rein is attached at the lower end have considerable leverage so very severe pressure can be exerted. A sensitive touch, with 'kind' hands, is therefore needed.

Before choosing your bit, ask the previous owner which type the horse goes best in, and even if you do not care for this type it might be wise to continue to use it until you get used to the horse and can then bring about introducing your own choice. Lastly, if you buy

Fitting the bit. (Below left) Incorrect: this bit is too narrow and is squeezing the horse's lips uncomfortably. (Centre) Incorrect: this bit is too wide and will slop from side to side; the horse will also be able to get his tongue over it. (Right) Correct: the bit is sitting comfortably against the side of the mouth, giving a sensitive 'feel' and allowing the correct aids to be given.

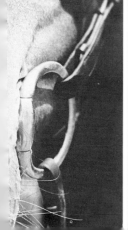
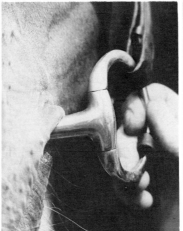
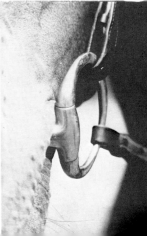

second-hand, make sure the joints and swivel surfaces are not badly worn as this will be likely to cause pinching.

Other items

Many other small items will be wanted; I will list the practical minimum that I consider necessary for those keeping a horse or pony for the first time. A stiff broom, pitchfork, shovel and a plastic or rubber bucket will be needed for stable duties, together with a galvanised bucket for making mashes. The grooming kit should consist of a hoofpick, soft body brush, curry comb, stiff dandy brush, a sponge, towel and cloth. For tack cleaning, a bar or tin of saddle soap, a leather dressing such as Hydrophane, a small sponge and two cloths will be required.

A first aid kit will also be necessary; a simple one to start off with should consist of a pair of surgical scissors, roll of cotton wool, calico bandage, tin of kaolin poultice, an antiseptic cream for small cuts and scratches, an ointment of the kind that helps prevent cracked heels and assists in the rapid healing of cuts and protects against bacterial infection.

Headcollars: (Below left) a leather headcollar of good strong construction that is functional as well as nice-looking; (Right) this nylon headcollar is cheaper than its leather counterpart but will nevertheless function well, and being more closely fitting will also be less easily slipped off by the horse.

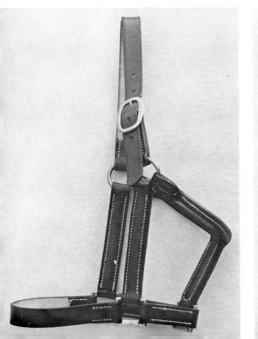
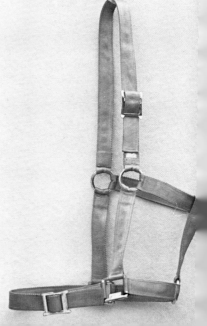

If you intend to use a haynet, don't waste money on the more expensive tarred variety, which are said to discourage a horse from chewing them. I have found this makes no difference – in fact I have one horse who seems to take a positive fancy to the tarry taste!

You will need a good, well-fitting headcollar. This is another item where money can be saved: nylon headcollars are cheap, strong, virtually maintenance-free and much less easy for the horse or pony to 'slip' than the more expensive leather ones. A lead rope is also a necessity, and here again nylon cannot be bettered. You will need a saddle rack for storing your saddle and a wooden horse for cleaning it on. I find it best not to make the wooden horse used for cleaning the saddle also serve as a rack for storing the saddle, as it is often easier to do the cleaning in the feed room and then to store the saddle in a more secure tack room or in the house. These stands can be obtained at a fairly reasonable price; that illustrated in the plate section, is particularly useful in that it can be adjusted to cradle the tree of the saddle and so allow for easy cleaning of the inside. A very simple but effective saddle stand and drying rack can also be made from odd pieces of wood and lengths of tubular steel. I have included a photograph of the one that I made from the metal parts of a collapsed fruit cage. Old metal conduit pipe of the type that was used to encase single electric cables and is now obsolete can also be adapted for this. Space these to provide hanging rails and slot them into holes drilled in the wooden uprights. The photograph will show you how this is assembled and fitted up, and you can construct your own with materials and space available.

Another saving easily made both in cost and in saddle cleaning is to use a home-made saddle cloth. These are best in cotton; my wife makes them from cutting up and stitching old sheets. They need to be big enough to cover the area of the horse's back and sides where the saddle fits; if used beneath the saddle when the horse is exercised they keep the saddle clean and also provide a friction-absorbing layer between the back of the animal and the panel of the saddle. I find there is no tendency for them to 'ruck up' and they are easily washed and dried.

If your horse or pony is to be wintered out with a New Zealand rug I have suggested earlier the benefit of having two rugs for alternate use. This is, of course, expensive, and you may decide that to do this will add too much to your initial outlay. Don't forget, though, that this piece of clothing will have to stand up to some very hard wear so have the best you can afford and also remember that one rug will get twice the wear of two, and will wear more quickly because of having less chance to dry out or be cleaned. If you are to provide your horse with a stable rug the type made from jute and lined with a wool-like material will be satisfactory, and probably the cheapest; a webbing surcingle to keep it in place is often sewn onto it. Personally I prefer one without a surcingle attached, which then requires a separate surcingle with a roller and felt pad to prevent pressure on the spine. If your funds will not allow for this it is possible to use an old numnah on top of the stable rug and under the surcingle which will help to prevent undue pressure where the surcingle goes across the horse's back.

Storage bins for concentrates

The best storage for grain and other concentrates is undoubtedly that provided by the galvanised bins manufactured for this purpose. They are vermin-proof, and although expensive will last a lifetime if placed somewhere dry and with the precaution of a piece of damp-proof material put underneath them to prevent dampness from the floor causing corrosion. I use them and find it best to store the crushed oats, etc., in their hessian or paper bags placed inside the bins so that each batch is fully used up without the risk of old or stale amounts being left in the bottom of the bin. If you decide to make your own storage bins and choose to do this from wood, make sure that it is strong and tight-fitting, and if possible line at least the bottom part with sheet metal to prevent vermin from chewing their way through. Plastic refuge or garbage bins are sometimes bought and kept for storing feed, but I have found that these are inclined to sweat, to the detriment of the feed they contain.

7 Stable routine and discipline

The activity that brings you in the closest contact with your horse or pony is, apart from riding, when you are grooming, and this will most probably be in the restricted confines of the stable. From the outset, therefore, you must insist that the animal accepts your presence and displays no bad manners towards you. Be kind but firm, and insist that he is obedient while being attended in his box. Apart from standing still while you move around him he must also learn to move over for you. This is easily achieved by placing a hand firmly on his rump; give a push in the desired direction so that he moves away from you, and at the same time give a word of command like 'Get over'. Very soon all you will need to do is step back and give the word of command and he will automatically move in the desired direction. I am a great believer in voice training; most horses will learn to respond to quite a few verbal instructions; by keeping to a regular routine and order of tasks performed your horse or pony will quickly learn the procedure and associate the commands with the necessary actions.

Daily inspections and worm control

Probably the most important single aspect of your horse or pony's welfare is regular inspection of and attention to the feet. Chapter 8 deals fully with hoof care and shoeing, so I will restrict my advice here to that of cleaning out a hoof that has become packed with dung or mud. Always use a proper hoofpick that is specially designed for the job and has a blunt point, and not a sharp pointed object such as a pair of scissors or a knife. Using the pick, begin by removing the dung or mud from inside the shoe at the heel, working towards the toe; work round both sides in this way and then gently clear each side of the frog working towards the point. This will ensure that you do not cause damage to the soft parts of the frog. Finally, clear the cleft of the frog and gently insert the pick at the base of the frog (the wide part at the heel) and remove the accumulated muck. This should be done carefully to ensure that you do not inadvertently prise up the shoe, especially when dealing with the hind feet. Tap the shoe to make sure that it is not loose before moving to the next foot.

Another daily check that should be carried out is to make sure that the animal is staling regularly and that droppings are both regular and of a consistent nature. If you notice that the horse has not staled during the day and cannot be encouraged to do so, walk him out and then return him to his box onto some fresh straw or other bedding material. A whistle and some rustling of the straw will then usually do the trick. If he fails to urinate in spite of trying to do so it indicates that something is wrong, and a vet should be consulted. The colour of a horse's urine is also said to be important in indicating general health but conflicting advice is often given about what colour is normal and what the different colours may indicate is wrong. Some will say that the urine should be colourless, others that it should be straw-coloured or light brown; yellow urine is said by some to indicate kidney trouble while others will say that a dark brown colour is the sign of this. In fact all of these can be true – or none of them. In practice the urine of many horses will alternate through all these colours without it indicating anything worse than a slight temporary upset. A change of diet, an increase or decrease in the amount of exercise, even the time of year, can all make a difference to the colour of the urine. Soluble waste products are eliminated from the body by two means: by sweating, through the pores of the skin, and by the kidneys and urination. In hot weather when sweating is increased less urine will be passed – and this will therefore sometimes be thicker or cloudy; in cold weather the pores will remain closed and the kidneys will increase their output so that more urine is passed. Another common cause for a change in the regular colour of the urine is a slight chill brought about by the animal being allowed to cool off too quickly after exercise. Provided that there are no other symptoms to indicate that something is amiss, variations in the colour of the urine can therefore be regarded as nothing more than the slight changes of physiology to which all living creatures are subject from time to time. If, however, you are worried about an unusual colour and this persists over a period of a month or two, then it is obviously best discussed with your vet.

The other poisonous material that the body needs to eliminate is the insoluble waste, with faeces by the bowel. With horses the common term for this is 'droppings', and here again the horse owner should make a daily check to see that regular evacuation is taking place. Droppings are produced by the action of involuntary muscles that squeeze the unwanted by-products of the digestive tract along the twisted tube of the gut to be discharged through the anus. Glandular structures lubricate the passage of this waste material and thus we should expect to see droppings that resemble a series of linked balls, very slightly oily and crumbly in texture, with a colour varying from light to dark brown. Several things can cause a variation to this pattern. If droppings become sloppy and resemble a cow-pat it is often a sign of too sudden a change in diet–from grass to stable feeding or vice versa–and they will probably also be a green colour. Sometimes a chill will produce sloppy droppings, and colour variations caused by changes in the type of concentrates being fed can also result, especially if the changes are not gradually introduced. On the other hand you may find that the droppings are very dry and hard-packed. Should this be the case it is likely that this constipation is being caused by the incorrect feeding of concentrated food without enough roughage. If no other symptoms–such as those described in chapter 5 for colic–are present, give a bran mash and then adopt the obvious remedy when feeding. Should the animal's droppings become loose, yellow or mustard in colour, and strong smelling when the diet has not been changed, this could indicate that the animal needs to be wormed as a build-up has taken place. There are several groups of worms that infest equines. Their presence in small numbers can be tolerated by the animal without harm, but a large build-up will lower the animal's general condition. The adult red worm or strongyle will attach itself to the gut wall and suck blood, while other types live on the food inside the gut. Regular worming will keep the infestation at an acceptable level; the frequency of the worming will depend on the conditions under which the animal is kept and the number of other horses sharing pastures, etc., because worm larvae are ingested by the horse or pony when feeding. If the horse or pony shares a small paddock area with several others treatments every six weeks will be required; if it is stabled or in a large, well-managed paddock worming every six months will do. Worming treatments nowadays are usually given by one of two methods: either paste is squirted onto the back of the horse's tongue with a special syringe, which is said by its advocates to ensure that the animal gets the full dose; or a powder is mixed with the feed, which is the method I favour as I consider it to be easier and just as effective. Whichever method you use, it is advisable to change the make of the treatment each time to ensure that the parasites do not build up a resistance to a particular brand. A final word about droppings that are seen to become suddenly loose for no reason is that most horses will show this symptom if they become nervous or frightened or if their regular routine is changed.

The need for punishment

Truly vicious horses or ponies are fortunately rare, and post-mortems on animals that have been put down for this reason often reveal brain abnormalities. However, horses, like humans, have their 'off' days when they are less sweet-tempered than usual. If the misbehaviour is only mild, be firm but speak in a gentle tone as it may only be due to nervousness on the part of the horse and the more calmly you behave the sooner the horse will gain confidence in you. If the horse is positively disobedient or becomes bossy, as some horses will with a new handler, it should be scolded in a harsh voice. If the bossiness extends to nipping you or, worse, to kicking, then show the animal a stick as you scold. If this doesn't put a stop to the bad habit, keep the stick handy and the next time it happens, use it. A short sharp reminder is all that's required. Do not beat the animal, and be sure that any punishment is given immediately the crime is committed – a whack given half a minute later will not be related by the animal to the offence. In the majority of cases, however, and assuming that you haven't bought a 'bad un', the animal will quickly get used to you and be obedient. By talking to him and letting him know of your presence and intentions, with firm but kind handling you should have no trouble. You may find that geldings are slightly better in this respect than mares, as the latter can be a little touchy at certain times and this has to be allowed for. Some mares can get rather nasty when coming into or going out of season, becoming extremely friendly when actually in heat. Others may act in reverse, and as there are no hard and fast rules you will need to be vigilant if you buy a mare until you get to know her habits. My horses are used to my being in their box to pick up a dropping or attend to something even when they are feeding, and I never have any trouble; naturally, though, at feeding time one wants to leave them as undisturbed as possible. Lastly, while tolerating no bad temper from your horse, be sure that there is none from you either, as this will only make things worse rather than better.

Grooming the stabled horse

Advice on how much grooming should be given each day varies between a full hour, of which half should be devoted to the animal being 'strapped' or 'banged' (these are terms given to the practice of applying vigorous massage to groups of muscles), to the need for only a quick brush over. Grooming requirements are generally likely to fall somewhere between these two extremes; the need also varies with the amount of work that the animal is doing. Much of the toxic waste that is produced by the working animal will be eliminated through the skin, and regular grooming that improves the skin tone and frees the pores that may be blocked will greatly assist this elimination. This is the reason for grooming a stabled horse, together with that of improving his appearance. Another important function that daily grooming provides is that you become accustomed to the feel of each part of the animal and will immediately recognize anything unusual by way of lumps, bumps or heat. This can be invaluable in the early treatment of troubles that do arise, as well as possibly saving you time and money and the animal from unnecessary discomfort if something has gone wrong. Properly carried out, grooming will therefore benefit the stabled horse, who can receive much more thorough grooming than one kept outside, where the necessity for natural oils to remain in the coat for weather protection in the winter outweighs other considerations.

Begin grooming by starting at the poll, and work down the neck and front legs and then along the body to the hindquarters and hind legs. Do the left (near) side of the horse first and then move to the right (off) side and repeat the procedure. Talk to the animal, making your presence known. Stand far enough back so that the full weight of your arm and body can be used, and always brush with long strokes along the direction of the hairs; it is often useful to balance yourself by placing your free hand on the horse. This also helps to let the animal know what you are doing and keeps him in position. Use the dandy brush first to remove dried mud and dung, but do not use it on the horse's head, especially round the eyes and mouth, nor to brush the mane or tail. Thin-skinned and clipped animals will be very sensitive to the use of this brush and in these cases its use should be limited to where hair and skin are thickest. It is often possible to get the animal to accept a brushing with the rubber or plastic type of curry comb (but *never* the metal kind) to remove dry mud and dung when he will not tolerate the prickly stiffness of the dandy brush. Next go completely over the horse or pony with the soft body brush in the same manner, cleaning it after every few strokes on the curry comb; finish up using this brush for the mane and tail. Use your fingers with the brush to remove tangles in the hair and brush well down into the roots to free dirt and scurf. Opinions differ as to what length manes and tails should be kept, but provided they are in good condition I prefer to see them rather long and full. This gives the horse maximum protection from the weather, and a good forelock and tail will help to keep off the flies in summer. Tails should be brushed out and kept trimmed off squarely at a point somewhere between the hocks and the fetlocks, depending on personal preference. Some breeds have good, strong tails that allow them to be kept long, but never let them become so long that they drag in the mud. Ragged manes that need to be shortened should have the unwanted hairs pulled and not cut; this will maintain their 'lay' and natural look. If you use a comb for the mane be careful not to drag it through the hairs and break them; I find that fingers and the body brush work best for tangled manes.

The horse's head should be groomed last, using the body brush for the forehead and cheeks but being very careful not to get too close to the eyes. Finally use a clean, dry cloth to wipe over the head and body to promote a shine on the coat. A useful piece of equipment that can be made for this is to cut out the pocket from an old pair of trousers or large jacket, stuff it with old wool or similar padding and then sew a soft chamois leather around it. This makes an excellent 'buffer' for putting the final touches to the coat of a groomed animal. It is sometimes advised to sponge out the nostrils of the horse and the area around the eyes. This will not harm the nostrils but I do not like to do this around the eyes as I do not consider it good practice to wet the very sensitive skin in this region and remove the delicate balance of oil in the skin here. I prefer gently to remove any grit or other matter that has collected in the corners by using a forefinger, and to brush and smooth the skin around the eye area with my fingertips.

Stains on the coat will have to be removed with a sponge, using soap and warm water. Dry the area afterwards before brushing. A method I often adopt is to do this at the time of the morning feed, so that when the horse is ready for grooming prior to exercise, these patches are dry and can be brushed over. Use a curry comb to loosen the build-up of dirt in the grooming brushes and give both curry comb and brush a sharp tap on the hard floor to remove these particles as you proceed. Keep your grooming equipment clean, and when washing brushes stand them on their bristles to drain and dry.

If your horse is a gelding his sheath should be seen to regularly; it will need to be washed out about once a week, more frequently than this in hot climates. I do this once a week as I feel that this interval

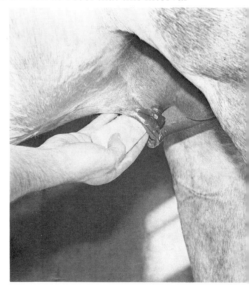

Regular cleaning of the sheath with geldings (and stallions) is necessary to maintain hygiene.
(Right) Cleaning inside the sheath with a small sponge dipped in warm water containing mild antiseptic.
(Below left) Pulling out the folds of skin to clean out the grease and dirt.
(Below right) Washing off the loose dried skin from the penis.

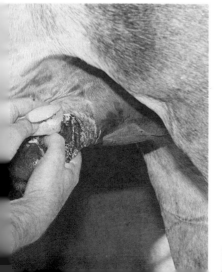

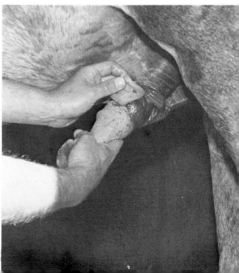

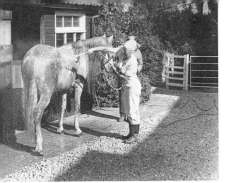 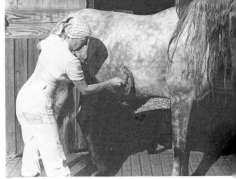

Above left: When hosing off your horse or pony begin by playing the jet of water onto his feet and lower legs to get him used to it, then work along his body from front to rear. Above right: Removing water from the coat with a sweat scraper, which is drawn over the body in the direction that the hairs lie; start at the top and work down and under the horse's body.

enables the natural oil in the skin to be maintained at a reasonable level. It is sometimes suggested that vaseline should be smeared onto the skin after washing, but this is not necessary if the interval is kept to several days and I think that it adds to the collecting of sticky deposits. Use warm water to which a little mild, non-stinging antiseptic has been added. Wet a small sponge in this and rub on some soap. Then gently insert this into and around the sheath and clean out the accumulated matter; the folds in the skin should be particularly attended to. You will be in a vulnerable position while doing this, and when first attempting it approach the horse slowly and speak reassuringly to him. Place one hand on his back and only use one hand with the sponge at first until you are sure the horse will accept it. Later you will find it necessary to use your free hand to pull open the skin folds to remove the grease and dirt.

Weather permitting, I also like to hose or wash off each horse occasionally and to shampoo mane and tail. Choose a windless day for this and always rinse out with clean water any shampoo or soapy liquid that is used. After washing, remove surplus water with a scraper (see illustration) and then allow the horse or pony to dry off in the sun or in an airy (but not draughty) box. If you turn him out before he is dry he will immediately roll and smother himself with grass sap and dirt, together with dung if there is any handy! When washing the animal–and especially if you use a hose–make sure that you do not get any water in his ears. Finally, if for any reason you cannot find enough time every day to groom the horse properly, do try to do this at least once a week and always make him at least presentable before riding out.

Cleaning and storing tack

Leather tack requires the most attention, though properly cared for it has the longest life. Its worst enemies are water and heat. Hot water will remove the natural fat content of the leather, and heat dries it out and causes it to crack. Always use a damp cloth or sponge to clean off mud, and then use one of the preparations that are specially sold for leather care. I do not think there is much difference between the various makes that can be obtained; it really amounts to trying a few and then making a personal choice of the one you like best. It should be noted that leather has a 'rough' side and a 'finished' side; oils intended to keep it supple should be rubbed into the porous surface of the rough side, while saddle soap should be used to put a sheen on the finished side. Detailed instructions about their use will be found on the tins containing the cleaning material, but generally a weekly application of oil should be enough, while the cleaning and polish with saddle soap will be necessary after each day's use. If you cannot find the time to dismantle your tack every day for cleaning, try to do the job properly once a week by stripping it down. This will ensure that buckle and strap parts are thoroughly cleaned and will also result in your giving the tack a proper check for worn parts, broken stitching or cracks. If you intend to store the tack for a fairly long period always take every part to pieces and clean and oil it thoroughly, give all the buckles a smear with grease and then reassemble everything: tack should be stored in a cool, dry room. Don't forget to check it over occasionally to see if any moisture that was present has caused the leather to turn

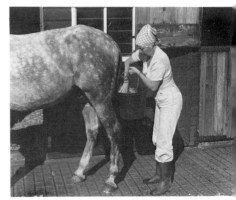

To wash the tail, give it a thorough dunking in a bucket half filled with warm water to which shampoo has been added; use a sponge to reach the upper part of the tail and the dock. Unless you know your horse very well, it is advisable to have somebody at his head to reassure him at first, as though it is safer to stand to one side in case he kicks it is then very difficult to wash the tail properly.

green with mildew. If it has, the remedy is a repeat of the same treatment you gave it before storing.

Items of tack other than leather, such as nylon or lampwick girths and the various types of numnah, will require regular washing to keep them in good order. They will often need to be scrubbed with soap and water to remove mud and sweat and then rinsed with clean water. When hanging girths up to dry tie or peg both ends to the line and let the middle part hang down in a loop. This will ensure that the buckles do not have the water draining through them, causing rust, and the leather ends and stitching do not become cracked or broken for the same reason. Bits need only to be washed in soapy water and rinsed and dried to keep them in perfect condition; check for sloppy joints and rings that could be causing them to pinch the animal's lips. The golden rule to remember when storing tack and equipment is to put it away in a state of instant readiness for when it is to be used again.

Keeping records

I once read a piece of advice that has always stuck in my mind and has proved to be most beneficial. It is 'plan your work–then work your plan'. If you follow this advice you will find it useful to include a blackboard in your feed and tack room so that the many tasks and items that have to be performed can be chalked up as a constant reminder. It is very easy to forget dates such as when your horse was last wormed or received its last injection against flu, while if you discipline yourself to use the blackboard it only takes a moment now and again to keep these things up to date. Never neglect to have your horse or pony given its jabs against tetanus and equine influenza, booster jabs for which should, according to recent opinion, be given every nine months. (Yearly intervals used to be thought sufficient.) I also chalk up the dates that each horse was last shod or given any special food or treatment, and its daily feed requirements. This is particularly helpful to anyone who wants to check in my absence, and allows for the correct feeding to be carried out when I am away and someone else takes over. Tasks carried out in a regular and orderly fashion will result in a better cared for horse or pony, cleaner and more hygienic stables and a happier and more efficient atmosphere.

8 Care of the feet

The feet are more exposed to injury and abuse than any other part of the horse. No matter how good an animal is in every other way, if its feet are faulty the rest counts for nothing. When one considers, therefore, how relatively little time and money needs to be spent in keeping the feet in good condition it is surprising how many horses and ponies fail in this respect through neglect. Some faults are congenital, and only breeders can play a part here in trying to eliminate weaknesses by selective breeding; however, many disorders can and do take place through bad management.

The foot is a very complex mechanism comprising sensitive interior structures encased in a layer of insensitive horn that forms the wall, sole and frog. The horny wall is a continuation of the skin, which changes to form a rim of horn at the coronary band. This horn is covered by a thin, extra-hard outer layer produced by the periople. This outer layer controls the moisture content of the hoof wall and prevents the wall from losing too much moisture and thus becoming brittle. The periople is situated immediately above the coronary band; both are found around the top edge of the hoof wall where this joins the hairline. Because the horn is produced from the skin it takes its colour from this–i.e. if the skin above the hoof is white, having no pigment, the horn will also be white. White horn is generally considered to be more brittle than the black or slate grey horn formed by dark-skinned animals, and many people prefer a horse possessing dark hooves for this reason. Don't be confused into thinking that white skin means white hairs–many grey horses have black skins; it is also possible to have a different coloured skin on the different legs of the same horse.

The insensitive wall of the foot is non-vascular, and should be thick enough for nails to be driven through when the horse is being

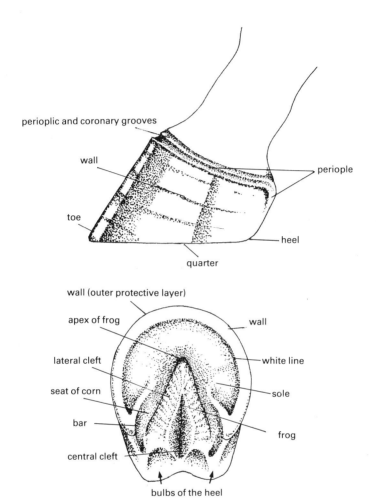

Fig. 5 A lateral view of the horse's foot, and its underside showing the bearing surface.

shod without the wall cracking or breaking. The sole should be hard, and thick enough to protect the sensitive inner structures above it. It should be the shape of an inverted saucer on the front feet and rather more deeply vaulted on the hind feet. Horses or ponies with shallow, rather flat soles are more likely to suffer from bruised soles and foot ailments caused by hard, uneven going, especially if the soles are thin and yield to pressure. Don't be alarmed if on inspection the sole appears to be flaking in scale-like pieces. A normal, healthy sole grows by producing flakes of horn that flake off when they have reached sufficient length.

The frog should be wedge-shaped and be of a texture similar to that of india-rubber. Its bearing surface should be in contact with the ground when the horse places his weight on the foot. The frog performs several important functions when this happens: it cushions the bones and the tendons of the foot; aids the circulation of the sensitive parts by the pumping action of the pressure exerted upon it; being softer than the rest of the hoof it provides the grip necessary for the horse's foothold. The frog will 'peel' or shed its outer layer as new growth occurs, and like the sole will thus regulate its size and formation. Only the wall of the foot continues to grow and is not self-regulating. If not in wear (as for instance, when shoes are worn) the wall needs to be trimmed back to maintain its correct angle and length.

Because of the necessity to protect the feet of the working horse or pony from excessive wear the method of shoeing as we know it today has been developed. Although this allows us to use the animal for purposes that would otherwise not be possible, the act of shoeing does at the same time create certain problems. There is the unnatural concussion to which the foot is subjected when the nailing of the shoe takes place; the weakening effect of the continual nailing through the horny wall; the drying effect of hot shoeing, which causes evaporation of the natural fluid around the lower edge of the hoof; loss of some of the protective outer layer of the wall through rasping, which causes loss of moisture and increased brittleness; restriction to the natural expansion of the foot by the confines of the shoe. The importance of using a qualified farrier who understands these factors and so minimizes them where he can will, therefore, be appreciated by the reader.

Shoeing

Farriery, properly carried out, is probably the most important single factor of your horse's care and well being. It may be no easy matter to find a blacksmith at all in some areas, let alone a good one, but the effort spent in seeking out the best man possible will more than repay your trouble. Whether you take your horse or pony to him or

A correctly fitted shoe of the concave, fullered type used for hunting and for everyday activities. This is a mild steel, machine-made shoe that has been shaped and fitted hot.

whether he comes to you will depend on your particular circumstances, but if you are lucky enough to have a good blacksmith nearby so that you can ride the animal to him then this will be cheaper. If the horse has thrown a shoe and is footsore, or if the shoes have become so badly worn that it is dangerous to ride, it will be necessary to arrange for transport for the animal if the farrier cannot come to you. For reasons of cost alone it is in your interest not to let things deteriorate this far, and if you consider the risk of injury to both yourself and the animal, to be this neglectful is just plain stupid.

Depending on the amount and type of work your horse or pony is doing and the ground conditions, shoes will need to be replaced at between four and seven weeks. I find that a good average is five weeks, and like to have the shoes removed after this interval in any case so that the feet may be trimmed. Shoes that are left on for more than seven weeks, even if the horse has only been doing light work and the shoes themselves are not badly worn, will nevertheless cause problems. The outer wall of the hoof will have grown wider than the shoe and the edges of the horn will become liable to damage by cracking and peeling back. With the growth of the foot the shoe will also have been carried forward, causing the heel of the shoe to lose contact with the bars (these form a vee between the sole and the frog and provide a hard ridge for the ends of the shoe to rest upon); instead, the shoe will cause pressure and bruising to the sole where it is in contact. This area of the sole that is sensitive to pressure from a badly fitting shoe is known as the seat of corn and is usually found on the front feet, which have shallower, less vaulted soles than the hind feet.

Two methods of shoeing are employed—'hot' and 'cold'. If carried out by a capable farrier, cold shoeing can be very good and has some advantages. There is no danger of a hot shoe charring the hoof too much when being fitted, as can sometimes happen. When it does, if this charred area is not rasped away before the shoe is nailed

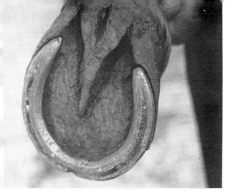 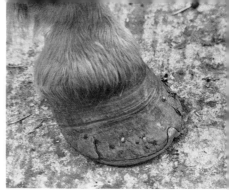

Above left: Although it is not badly worn, this shoe has nevertheless been left on too long. Note how the outside wall of the foot is growing down over the shoe. Above right: This shoe has also been left on too long, resulting in the wall of the hoof overgrowing at the quarters, cracking and peeling back. The clenches (protruding nail-ends) are also raised, which can cause the animal to cut itself.

on, in wear the burnt horn beneath crumbles away and a loose shoe results. The slackness of the shoe will be seen by the raised clenches around the outer wall of the hoof. When properly made, these clenches should lie neatly around the outer wall, as they are made from the broken off nails that are turned over by the blacksmith as he tightens the shoe after nailing it on. With cold shoeing there is also no danger of the lower rim of the hoof wall drying out through heat being applied. On the other hand, when properly carried out hot shoeing makes it easier for the blacksmith to get an even fit between hoof and shoe. Because the hoof softens and expands slightly during this heating it makes it easier to knock the nails through and when the horn cools and contracts it will grip round the nails, giving them a firmer hold. I normally have my horses shod hot, but I ask the farrier to keep the burning to a minimum. In bad cases of brittle feet I prefer cold shoeing. Once again, though, you are at the mercy of the man in your area, and if there is not another blacksmith near enough that you can go to, you will have to use some tact and persuasion to get the desired result.

For normal purposes there is a standard type of shoe, which is nowadays usually machine-made from mild steel. Shoes are produced in different sizes, and the blacksmith makes the final adjustments after selecting the nearest fit. For horses or ponies used for activities that require extra grip, such as when jumping or turning sharply, shoes fitted with special studs can be obtained. Studded shoes are also used in some countries to enable the horse to work in icy conditions where normal shoes would not provide sufficient grip.

Corrective shoeing is another service that can be provided by the expert farrier. If minor defects in the conformation of the legs produce a faulty action, this can often be alleviated by attention to the horse's feet. This is particularly true with defects in young stock and corrective measures to alter the uneven distribution of weight can result in considerable improvement to the defect. Corrective treatment can also be applied to relieve a condition known as sandcracks. These cracks in the wall of the hoof originate at the coronet and extend down the wall; various measures can be applied by the farrier to control the spread and length of the crack.

Going without shoes

There are times when a horse or pony can be left unshod, and in some circumstances it can be beneficial. In some hot countries where the going is consistently that of hard sand, horses are sometimes worked without shoes but the type of animal used has feet with very hard, concave soles. Generally speaking, unshod horses should not be worked unless the work is of a very light nature on easy going – working in an indoor school, for instance. When a horse or pony is turned out for a period of rest, having its shoes removed will ensure that if other horses are also kept in the same field any kicking will be less likely to cause injury. The unshod hoof will also be given chance to expand slightly without the restriction of a nailed-on shoe, and the horn will also be able to grow down without the weakening effect of the nail holes. To prevent cracking and chipping, especially to the front feet, it is sometimes advisable to have grass tips fitted. These are small half-shoes that prevent the toe from becoming damaged. It will be necessary to inspect your horse or pony's feet regularly while he is being rested at grass; every four to six weeks the farrier should trim the feet, and where grass tips are fitted these will have to be removed so that the overgrown horn can be trimmed back.

Your horse will also be without a shoe if it is 'thrown' through being allowed to wear down too much. If you are riding at the time you will often hear or see this happen, but if not you should dismount the moment that you notice something wrong with the animal's gait or balance. Inspect the foot to make sure it is not damaged, and if all is well you can ride quietly home, keeping to fields or grass verges and dismounting on stony going or metalled roads. If you fail to do this the chances are that your horse will suffer

Picking up a foreleg. (Above left) Stand by the horse's shoulder facing to the rear; run your hand gently but firmly down the back of the leg. (Centre) When your hand reaches the fetlock grasp the pastern below it and ask for the foot by applying an upward pressure. (Right) As the foot is lifted up, step forward and support its weight.

from a sore foot and will then have to be rested until it is better. Whether you are able to ride the animal home or have to walk him, remove any projecting nails that may have been left in the foot, as these can cause injury to the other legs.

Daily hoof care

The working horse or pony should have its feet inspected every morning and again after exercise. In order to make your daily inspection and to carry out the necessary attention to the hooves you will need to pick up the animal's feet, and he should be handled in this way without fuss. Indeed, many horses and ponies will become so used to the routine that they will automatically lift up their front foot when you approach with the hoofpick. Always begin with the near foreleg. Stand at the animal's shoulder facing the rear, run your hand down the back of the leg and grasp the pastern; ask for the foot and apply an upwards pressure, whereupon the animal should willingly lift it up for you. Next go to the near hind, and this time work your hand slowly but firmly across his quarters and down the rear of the leg to a point between the hock and the fetlock joint. From here you ask for the foot by pulling the leg forwards and as soon as the animal lifts its leg you step in and support the foot with your free hand. The off side is then done.

Regular inspections carried out in this way will enable you to see that there are no raised clenches that can cause a cut by the animal 'brushing' itself or when getting up after resting or rolling. (Brushing is the term used when a horse strikes the fetlock of one leg with the foot of the leg opposite to it. It can happen with tired or unfit horses as well as with those possessing a faulty action.) While you

Picking up a hind leg – the usual method. (Above left) Stand at the horse's side and run your nearest hand smoothly down the quarters and on down the back of the leg. (Centre) When you reach a point midway between the hock and the fetlock exert a forward and upward pressure to ask the horse to lift its foot. (Right) As the foot is lifted up, step forward and support its weight with your other hand.

are holding the foot up any stones present must be removed with a hoofpick, and also any dung or dirt (see chapter 7). If this is not removed frequently the foot will probably become infected with thrush. This is a condition affecting the frog and is mainly, though by no means always, found in the front feet. The cleft and lower

Picking up a hind leg – the author's method. (Below left) As the hand drops below the hock move it round so that the thumb and forefinger run down the front of the lower leg. (Centre) Grasp the front lower end of the cannon bone between thumb and forefinger and pull forwards and upwards to ask the horse to lift its foot. This method is preferred by the author because if the horse snatches its leg up it will push the leg more firmly into your hand, whereas with the usual method you will probably have to let go; also, if you grip the leg in this way your arm will automatically stiffen and the movement of the horse's leg will travel through your arm and be counteracted by it, pushing your body away from the kick. (Right) Once the foot has been lifted, step forwards and support it with your other hand in the usual way.

sides of the frog become diseased; there is no mistaking the condition as there is always an offensive smell, though lameness does not result unless it is allowed to progress to an advanced stage. If thrush is found the decomposing matter should be picked out and the underside of the foot scrubbed with a stiff brush using water and an antiseptic disinfectant. This treatment should be carried out daily until the foot returns to normal. Regular inspection of the feet will also show you that the shoes are wearing evenly, without any difference in the rate of wear between one side of the shoe and the other. Uneven wear may be due either to faulty conformation of your horse or because of the incorrect fitting of the shoe. Discuss this problem with your farrier if it does occur, to determine the cause and see what corrective measures can be taken. If your animal has flat feet with very shallow, concave soles that become easily bruised you may have to resort to having leather protective pads fitted under the shoes. This is another problem that your farrier can advise and act upon. Personally I do not like them as one cannot see what is going on underneath and dirt, grit and dung often accumulate.

With the problem of brittle hooves, the answer is often said to be not buying an animal with brittle hooves in the first place! However, unless this condition is very marked, it is not in itself a good enough reason for rejecting an otherwise good animal. Quite a lot can be done to improve hooves that tend to become brittle, but there are also things that will increase and aggravate the condition. Hot dry weather, causing loss of moisture in the hoof, will lead to increased brittleness, and the answer here is to wash or soak the feet in water. Bad shoeing practices should also be avoided: excessive rasping down of the toe (called 'dumping'); rasping off the overlapping wall to make the hoof fit a shoe that is too small; removing too much of the protective outer layer of the wall when finishing off the clenches. If your blacksmith is guilty of this and cannot be persuaded to refrain from these habits while working on your horse, look for someone else.

To help seal the hoof and prevent loss of its natural moisture various types of hoof oil and grease can be obtained. These are applied to the outer wall as a protective measure and certainly do help. Some are claimed to soak in and help to make the horn softer and more pliable, but their effectiveness in doing this is in my opinion open to some doubt. You may also wish to increase the rate of growth of the horn in order to get any damaged parts grown out.

A hind shoe that has been left on far too long and is now almost coming off. The wear is uneven, indicating a faulty action that may be due either to defects or perhaps to a foot badly dressed by the farrier.

Again, there are products that claim to do this if applied to the rim of the hoof where the hair and skin join the foot at the perioplic groove and coronary band. The treatment must be diligently carried out for several months before any worthwhile results are seen, and in my experience the degree of improvement varies greatly from one horse to another. Apart from speeding up the growth in order to get cracked or broken parts grown out, one also wants to promote an improvement in the quality of the new horn. Several food additives claim to do this. Once again the desired result, whether marked or only marginal, has to be waited for and the treatment should be given several months' trial. Fairly recent research has resulted in the manufacture of products made from powdered seaweed that have given encouraging results in improving horn quality.

Finally, in your daily examinations of the feet you should notice any heat, and in particular any difference between the heat of each hoof. Heat should always be regarded with suspicion and as a likely sign of trouble, though after standing on deep litter and after

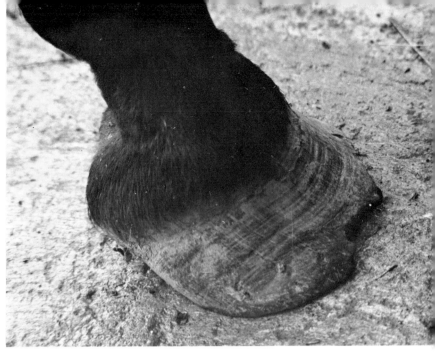

A neglected foot: the shoe has been lost and the horn at the toe has broken away badly.

The damaged side wall of a hind foot in need of reshoeing.

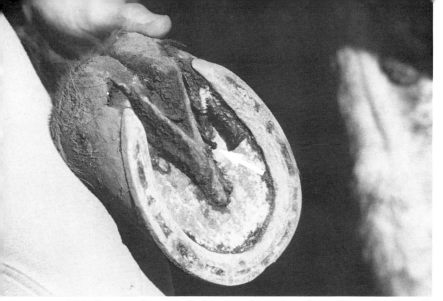

This foot was inspected after exercise, when the nail was discovered in the sole of the foot. Luckily it had lodged in the thick part at the bar and no serious damage or lameness resulted.

exercise the feet can be naturally slightly warm, as they often are for a short time just after shoeing. If this is very marked, however, and continues for more than a day, or if there is a great difference in the heat of one foot compared with the others, then one should look for the cause. Check the horse for lameness. If he is lame and shoeing has recently been carried out, have the blacksmith remove the shoe of that foot and check to see whether he has pricked the sensitive part of the foot. Even a nail driven close to the sensitive structure will sometimes cause enough pressure to give trouble. Alternatively, the sole of the foot may have been punctured by a nail or sharp object that has been picked up. In this case, once the offending object has been removed the puncture will have to be opened and allowed to drain. The foot should be soaked for half an hour in a bucket of salt water, and then a poultice must be applied to draw out the infection. Wounds of this kind are particularly susceptible to tetanus infection, and an injection against this should be given irrespective of whether the animal has previously been inoculated. If you are in doubt about whether or not your blacksmith and/or vet should be called because of suspected trouble in the foot of your animal, remember that old saying quoted earlier, 'no foot–no horse'.

Index

Age, 11, 12
American barn, 31
Bedding, 35–39
  deep litter, 28, 35, 38
  peat, 36
  shavings, 35
  straw, 37, 38
Bits, 76–79, 92
Blankets, 42
Boiled barley, 64, 65
'Bone', 3, 5
'Bowed hocks', 9
Branding, 49
Bran mash, 54, 65, 66
Bridles, 74, 76, 78
Brittle hooves, 95, 101, 102
Cannon bone, conformation, 5–8
Casting, prevention of, 40
Colic, 67, 68
Concentrates, 63, 64
Conformation, general, 3–10
Constipation, 85
Conversions, stable, 31
Corrective shoeing, 98
Costing, feed and bedding, 17, 18
'Cow-hocks', 9
Cubes, nuts, 64
Damp-course, 32
Day and night boxes, 38
Dealers, buying from, 15–17
Digestion, 59, 60, 63, 64, 67
Doors, stable, 26, 27, 41
Drainage, stable, 22, 25, 31
Draughts, stable, 26, 41
Drinkers, 28, 33

Droppings, 85, 86
Electricity, stable, 28, 30
Equipment, items of, 80, 81
Feed, storage of, 82
Feed store, 34
Feeding, amounts, 60
  for fitness, 54, 55
  of concentrates, 63, 64, 85
  of grains, 63
Feet, cleaning of, 83, 84
  trimming, 98
Fencing, 50–52
Fetlock joint, 8
Field shelter, 52
Fitness, working up to, 54
'Filled legs', 63
Fittings, stable, 28–30
Flooring, 24, 25, 32
Foreleg, conformation of, 6
Frog, 95–97
Galling, 73
Girths, types of, 73, 74
Grasses, types of, 47, 48
Grazing, problems, 46–49
Grooming, 83, 87
Handling, legs and feet, 99, 100
  punishment, 86
Hay, how to judge, 61–63
  storing, 32, 33
  types of, 61, 62
Haynets, 81
Hay-racks, fitting of, 28
Head, conformation of, 10
Headcollars, 80
Hind legs, conformation of, 7

Hocks, conformation of, 8, 9
Hoof care, 83, 84, 99–104
Hoof oil, 101
Hooves, heat, 102, 104
Horn, growth of, 101, 102
Hosing off, 90
Injections, anti-flu, 92
　anti-tetanus, 92
Laminitis, 46
Legal aspects, 20
Lighting, stable, 28
Loose-box, assembly, 32
　sizes, 20
Manure heaps, siting of, 40
Materials, building, 22–24
Mildew, dangers of, 33, 61, 63
Neck, conformation of, 10
Paddocks, care of, 48–50, 55
Pastern, conformation of, 8
Periople, 93, 102
Poisonous plants, 47
Reins, types of, 76
Roofing, 23, 24, 31
Rugs, New Zealand, 53
Saddle cloths, 81
Saddle fitting, 9, 70, 71
Saddles, types of, 69, 70
Salt, 67
Sectional buildings, 32
Severe weather, 58
Sheath, cleaning the, 89, 90
Shoeing, 95–98
Shoes, uneven wear, 101

Shoulder, conformation of, 9
Sick horse, feeding a, 66
'Sickle hocks', 8
Stables, building of, 20–31
　useful tips, 43, 44
Stains, cleaning of, 88, 89
Staling, 43, 44, 84
Stirrups, correct size of, 71, 72
Straw, mouldy, 37
　storing of, 32, 33
Studded shoes, 97, 98
Surcingle, use of a, 42
Tack, cleaning of, 91, 92
Tack room, 34
Tendons, 8
Tetanus, 92, 104
Threshold, tidy, 41
Thrush, 100, 101
Tie-rings, 28
Turning out to grass, 54, 55
Types, choosing, 1, 2, 10–14
Urination, 84
Ventilation, 25, 41
Vitamins, supplements, 67
Warranty, 15, 16
Water piping, 33
Water, provision of, 49, 67
　stable, 33
Weather protection, 52–54
Weight-carrying, 5, 12–14
Windows, stable, 26
Withers, conformation of, 9
Worm control, 55–57, 85, 86